Invitation to
The North

Hermann Tietgen

Marcia Tietgen Smith
and
Karen Smith Cade,
Editors

 FriesenPress

One Printers Way
Altona, MB R0G 0B0
Canada

www.friesenpress.com

Copyright © 2023 by Hermann Tietgen

Marcia Tietgen Smith and Karen Smith Cade, Editors

First Edition — 2023

All rights reserved.

No part of this publication may be reproduced in any form, or by any means, electronic or mechanical, including photocopying, recording, or any information browsing, storage, or retrieval system, without permission in writing from FriesenPress.

ISBN
978-1-03-914225-1 (Hardcover)
978-1-03-914224-4 (Paperback)
978-1-03-914226-8 (eBook)

1. BIOGRAPHY & AUTOBIOGRAPHY, PERSONAL MEMOIRS

Distributed to the trade by The Ingram Book Company

To Sarah, Rod + Wil,
happy reading always,
Marci Smith

CONTENTS

CHAPTER I   BLUE SUNDAY -BRIGHT MONDAY 1

CHAPTER II   WESTWARD FLIGHT 6

CHAPTER III   TO THE END OF THE STEEL 13

CHAPTER IV   NORTHWARD 18

CHAPTER V   FIRST CAMP 21

CHAPTER VI   FIRST STRIKE 29

CHAPTER VII   WHILE THE CAT IS AWAY 39

CHAPTER VIII WINTER CAMP AND FREEZE UP 53

CHAPTER IX   GOLDRUSH 67

CHAPTER X   DEEP WINTER 74

CHAPTER XI   REALLY COLD 77

CHAPTER XII   VACATION TRIP 83

CHAPTER XIII BACK TO THE BUSH 96

CHAPTER XIV   LOSING A FRIEND 100

CHAPTER XV   PROSPECTING AGAIN 102

CHAPTER XVI   CRACKING ROCK AGAIN 105

CHAPTER XVII SEASON'S END 115

CHAPTER XVIII FREEZE-UP 121

CHAPTER XIX   WINTER AGAIN 126

CHAPTER XX   DRILLING CAMP 130

CHAPTER XXI   STAKING TRIP 135

CHAPTER XXII JUST LIKE GOING HOME 140

CHAPTER XXIII	END OF DRILLING	148
CHAPTER XXIV	BREAK-UP	150
CHAPTER XXV	SUMMERTIME	160
CHAPTER XXVI	WEDDING TRIP	164
CHAPTER XXVII	A NEW LIFE	167
CHAPTER XXVIII	LIGHTS AND SHADOWS	171
CHAPTER XXIX	CITY LIFE	182
CHAPTER XXX	DEEPER INTO THE NORTH	188
CHAPTER XXXI	WAITING FOR SPRING	192
CHAPTER XXXII	AT THE OLD GAME AGAIN	196
CHAPTER XXXIII	ON THE GO AGAIN	201
CHAPTER XXXIV	OPEN WATER	209
CHAPTER XXXV	NEW CAMP	211
CHAPTER XXXVI	BABY NEEDS SHOES	216
CHAPTER XXXVII	ON THE MOVE AGAIN	222
CHAPTER XXXVIII	FREE GOLD	224
CHAPTER XXXIX	YELLOWKNIFE	232
CHAPTER XL	ITCHY FEET	237
CHAPTER XLI	FALL OF THE YEAR	243
CHAPTER XLII	LAST CAMP OF THE SEASON	248
CHAPTER XLIII	FLIGHT TO THE SOUTH	255
CHAPTER XLIV	WINTER IN TOWN	260
CHAPTER XLV	NORTHWARD AGAIN	263
CHAPTER XLVI	ASSESSMENT WORK	266
CHAPTER XLVII	A LOT OF SMOKE	275

CHAPTER I
BLUE SUNDAY - BRIGHT MONDAY

HERE IS MY story. It began in Montreal, in the month of May of the year 1935. I'd come to Canada 5 years ago, emigrating from Germany. This time around, I had spent three weeks in Canada's Metropolis, then still in the grip of the desperate depression. I was vainly searching for another job. Almost any kind of a job would have suited me, as long as it meant a chance to break away from that little village in the Laurentian Mountains where I had worked for the previous four years, just making a living, looking after some well-to-do men's country places. Plenty of skiing in the wintertime, and the company of charming friends on my days off, often helped to make me forget the fact that it was just filling the gap, and there had been no prospects of getting ahead.

Day after day I had tried my luck. But nothing would click. Promises came to naught, prospects faded out, and my few hard-earned dollars were dwindling only too quickly. *Before your last pennies are gone, you'd better get back to the mountains*, I told myself. I felt pretty blue that last day in Montreal. It was Sunday, and it was hot as I walked through almost deserted streets to the waterfront.

There was the mighty St. Lawrence River swirling and eddying around Victoria Pier, going, going to the sea, and with it went my thoughts, across the Atlantic and home. If I could only go… The harbour wasn't the right place for me. I wandered back up town. On Phillips Square bus drivers were peddling tickets. "Trips to Nowhere," I heard. I had to grin; "Nowhere is good. Where am I going?" I bought a ticket. Might as well find out. You know where that bus went? Half-way to the place I had tried so hard to get away from. It was as if someone were grinning behind my back. However, we got back to Phillips Square, having seen nothing new. *Even tomorrow will be far too early to go back to the Mountains*, I grumbled to myself, somewhat bitterly.

That night I had a few beers with a friend of mine.

We talked.

"There must be something a fellow can do, to get out of this rut," but there was the everlasting depression. Another fellow dropped in; he was one of the countless unemployed.

"You are lucky, Hermann," he told me. "You know at least where your next meal is coming from."

No doubt, he had a right to say so, nevertheless I couldn't be content. I was always hoping for a break. I didn't come to Canada just to make a bare living. "Let's have another beer. Here is to a better future, damn the depression."

The following morning, I went to see the boss to tell him I was going back to the old job at the cabin in the Laurentians north of Montreal. What to do with the last few hours in Montreal? Go and see friend Frank? My old chum from the Mountains was home, just having a belated breakfast.

It was about 11 a.m. "Been on a bender, I suppose?" I asked.

"Oh, no, but we had a swell party. You should have been with us," he replied.

"Well, nobody asked me," I replied. "I certainly had a great time all by myself."

"What did you do?"

"I just took a trip to 'Nowhere'," I said.

"Come on now. Tell me another one."

"That's the fact though," I sighed.

"Boy, you must have been desperate. What are your plans now?"

"I am going back to the Mountains. Things didn't pan out."

At this moment there entered another fellow who was shortish, broad shouldered, looking kind of familiar, but I couldn't place him.

Frank did the honours, "Pat, meet Hermann, an old friend of mine from the Mountains." We shook hands and from then on things happened fast.

"You want to go up North with me?" Pat asked without preliminaries.

"Sure. I'll go," I said.

It didn't occur to me to ask any questions. Somehow, I knew that this was the break I had been waiting and working for.

"Do you have to go back to the Laurentians first?" was Pat's next question.

"No, that won't be necessary. All my belongings are safely stored with friends. If I need anything, I have still got a few bucks left."

"That's fine then. We are going shopping this morning. Come along if you care. I've got the car in the lane. Maud will be ready in a minute."

Raven-haired, attractive Maud, smartly turned out, appeared on the scene.

"Maud, meet Hermann. He is the fellow who will go north with me." The car? It turned out to be an open Buick, 1924 or thereabouts vintage. The paint was slightly faded but the heart was still all right. We chug-chugged around town for a while. Then, "how about lunch?" said Pat.

"Fine with me."

Pat brought the car to a stop in front of the Lasalle Hotel.

"Care for a drink, Chum?" said Pat.

"Okay with me. What's it to be?"

"Hot Gin."

"On a day like this? Well, all right, make it two."

We had a few of them.

By and by, through our meal, Pat told me what little he knew a little more about the undertaking to which I had committed myself so abruptly. He had received an offer to take a prospecting party into the North.

He described, "Exactly where to I don't know yet, but we should soon learn. I am a pilot. In the wire I received I was told to meet the boss of this new outfit, I don't even know the name: company or syndicate or something, tomorrow morning at Toronto. I have wired my old mechanic to find out if he wants to go along, and then there is you if you still want to come."

"Of course, I want to come. You don't know how badly I need a change."

"Are you sure now?"

"Absolutely."

"Alright, I tell you what we'll do. I am leaving for Toronto tonight and will tell the boss about you. If it sounds okay to him, I'll wire you. Or wait, here is a better scheme, do you want to gamble?"

"Alright, shoot!"

"You pay your way to Toronto, and if you are not taken on, I'll pay your fare back. How does that sound to you?"

"Sounds fair enough, I accept."

"Tonight then, see you at the Bonaventure Train Station."

After the hot gin, the sun was somewhat hard on my eyes for a few minutes, but there was a new springiness in my step as I walked along Sainte Catherine Street, and that had nothing to do with the gin. A new life had begun, I had no doubt about that. At the train station I wrote a few postcards to the folks at home, "I am heading North, don't expect much mail for a while," and by phone I checked out with my Montreal friends.

There followed a slightly sentimental dinner at the Queens and then it was time to catch the train. I took it as a good omen that a somewhat dubious-looking character sold me a strictly non-transferable return half to Toronto at a cut-price. May the Canadian National Railway (CN) forgive me!

Breakfast at the Union Station in Toronto, and then to the Royal York Hotel to see the boss. We shook hands. The important man sort of looked me up and down.

"So, you want to go with Pat?"

"Yes sir. I would like to. But I was told it depended on you."

A friendly grin appeared on his face, "I think you will do. Good luck to you."

I was happy. Finally, life had taken on a new light, new horizons loomed. This, I felt, was one of those "Moments that have to be taken at the tide."

It was nearing lunchtime when all arrangements had been settled with the boss. In the lobby we met Charlie, an old friend of Pat's, and also a bush-pilot. Through our lunch he told us about the North and gave good advice concerning the supplies to take into the bush, which, in part at least, we intended to buy in Toronto.

"If you have nothing better to do, come along and have a look at the 'crate' they bought for us, Charlie," said Pat. At the de Havilland factory we found the crate to be a trim looking Fox Moth biplane, with a cabin for three between the wings, and the cockpit behind the wings in the fuselage. It would take a few more days to put the final touches to it before it could be brought over to the seaplane base for the mounting of the floats. So back to town we went and started on some serious shopping: Sleeping bags, tents, axes, iron rations, pots, and pans and so on, adding up the weight until we had our payload together, including first aid supplies.

Charlie was giving good advice, "You'll need plenty of mosquito dope! Those mosquitoes!" We had heard enough about them to frighten the bravest.

Sometimes you can hardly see the sun, and then the bulldogs. Why, they simply bite a chunk out of you and fly to the next tree to chew it up.

"Just to satisfy you, Charlie" Pat turned to the beaming druggist, "throw in another six bottles of Pine tar and Citronella, and have you got any Crazy Water crystals?"

"What do you want those for?" asked Pat.

"Well, now, a fellow might get blocked up!" said Charlie.

"You'll thank the Lord for them when you get constipated in the bush, what with the flies and mosquitoes around," cracked Charlie without blinking an eye.

Then we tried to get some maps of the country we were heading for, the district North of Lake Athabasca, where just a year ago gold had been found. We were going to see if we could find gold too.

Here, in Toronto, we were out of luck with respect to maps of the North. We were advised to get maps directly from Ottawa, and as we had to wait for our plane anyway, Pat decided to go to the Capital. But Toronto was new to me, so I went exploring.

CHAPTER II
WESTWARD FLIGHT

ON A FRIDAY morning our Fox Moth was finally brought down to the Toronto Seaplane Base; by evening the pontoons had been attached, and Saturday morning the test flights were undertaken. Then we loaded and gassed up our ship. We were ready for the first leg of our long flight to Edmonton and from there to the North. The big boss and some friends appeared to see us off. Pat climbed into the cockpit, I squeezed into the cabin, crowded even without me.

Charlie stepped onto the left float to crank the prop.

"Contact," Pat said to confirm that the ignition switch was now on.

"Contact," Charlie replied an understanding that pulling the prop would start the engine.

An expert twist, hammering and coughing the motor came to life, then settled down to an even roar. A last wave to our friends, Charlie gave a push and Pat taxied into the wind. The water was quite rough. Faster and faster, we skimmed over the waves. Sunshine painted tiny rainbows in the propeller-caught spray. The nose came up. We were on the step. We were off. Climbing fast, Pat circled the base, tipping his wings as a final salute to our friends below, and picked up our course slightly north of west.

Sudbury was our first goal. For a few minutes we were bumped around by the turbulent air over the big city of Toronto. Soon we were high above the white-capped waters of Georgian Bay, slowly the shadow of our plane crept along the course. The airspeed indicator on the wing showed that we were bucking a stiff headwind. Evening began to fall when finally the enormous stacks of the International Nickel Company, and Sudbury came in sight. On a lake near the town Pat brought our plane to a smooth landing and taxied toward a somewhat primitive dock. Rope in hand, I crawled from the cabin to the float, ready to ready to jump to shore. I was still pulling hard when a little man sidled up to me.

"Hey," he said out of the corner of his mouth. "does he take a drop?"

I looked at him, wondering why he should act so chummy. Again, he asked, "Does he take a drop?"

Pat had climbed ashore meanwhile, and out of the side pocket of the little man's overcoat appeared a nondescript bottle filled with a cloudy concoction. Moonshine, no doubt. With a grandiose flourish it was proffered first to Pat, then to me. Firewater! It almost brought the tears to our eyes.

As it turned out, the little man owned the dock. Surely, we were as welcome as the flowers in May! Wasn't he the greatest admirer of aviation and bush-pilots in all Sudbury? We were his guests, no getting away from that.

We had to meet all his friends. The town was ours; he would show us around. First though, we had to go up to the house to meet his folks, father-in-law, wife, brother-in-law and so on.

Thus, a memorable party began. The old bottle was passed around, a car appeared from somewhere, the whole gang piled in, and off we went downtown. A hotel room, beer, and some more moonshine. Toasts were drunk, speeches were started and never finished. Father-in-law offered me his farm in case our northern venture shouldn't pan out. The French temperament broke out more and more. Cheeks had to be kissed, tears in the eyes. *A votre santé!* Too your good health. *Bonne chance!* Good luck! Life is so beautiful, but, ah, so sad! *Alouette*, and *l'amour, toujours l'amour!* Love, always love!

Somebody had a bright idea: how about some dinner? *Mais oui!* But yes! Someone knew just the place. Into the car, out of the car. A Chinese food joint. Chop Suey. More friends, more booze.

"We must get out of this, it's almost midnight," Pat whispered to me. "We'll try to take off at 5 o'clock tomorrow morning."

Back to the hotel, Pat disappeared, and I talked father-in-law into taking me down to our plane. As well as possible I unrolled my eiderdown on top of our load.

"I'll call you in good time for breakfast," says beau-père. "Goodnight!"

"Goodnight! And thanks for everything!"

I must have fallen asleep immediately. It seemed only minutes later that I heard someone calling. "Armand, Armand, breakfast is ready." My real name, Hermann, translated well into French.

I rubbed my eyes, crawled out of the cabin, and cooled my roaring head in the lake. At the house, father-in-law, half pint, and a third fellow were assembled in the kitchen, balancing glasses, still or again? Somewhere an alarm clock was ticking. I glanced at it. 2 o'clock in the morning! Well, of all the .... It will be a long, long day.

The old man got busy at the stove. Oh yes, one had to eat. Soon some eggs, swimming in grease were making eyes at me. I never felt less appetite for eggs.

"Another drink might help," somebody suggested. I think it was the third fellow. He was a jail guard; I'd learned in the meantime. There was no getting out of it. Slowly the hours passed. I dozed off and on, sitting on a hard kitchen chair. Finally, the old man went off to fetch Pat. I heaved a sigh of relief. We would soon be out of this place.

Pat appeared, fresh as a daisy. A lengthy scene, goodbyes over and over again.

"For sure, we'll drop in again. Maybe in the fall. Who knows?" Pat wasn't sure.

With the sun at our back we were winging our way westward. Our course lay along the rocky North shore of Lake Superior toward Port Arthur. Down below a stiff southwest wind drove mighty combers crashing against the rocks. On and on the flight went, the motor thundering its monotonous melody. Again, and again my eyes closed, there is an uncertain feeling in my stomach. Then I must have fallen asleep. I came to with a start, and before I even knew where was, I was sacrificing to the gods of aviation in the most miserable aftermath of any party I've ever been to. While I was still trying to regain my composure, I noticed that our plane was going in a circle. Port Arthur? No. Looking out, I spotted three fires burning brightly in a triangle on the west side of a small roundish bay.

Pat's voice came through the speaking tube, "Going to land. Somebody in trouble down there."

Another turn and I saw a small figure near the shore waving frantically up to us.

At that time, I didn't know anything about flying an airplane, but even then, I well understood that it wouldn't be easy to land in that little bay. It looked so desperately small, and the main lake was rougher than ever. One seemed to hear the waves pounding against the shore as we were gliding down. Just inside the narrow inlet we hit the fairly smooth water of the bay, and a moment later our pontoons ground onto the sand of the beach.

The fellow who had been waving to us, came running.

"Is anything wrong?" Pat asked him.

"Nothing wrong. No. Nothing at all."

"Why in hell the three fires then?" snapped Pat.

"Just clearing some land. That's all."

Pat turned away, scowling, "The old beggar is bushed."

"Bushed?"

"Yea, too long in the bush by himself. Cracked, you know."

I got busy cleaning up the cabin after the catastrophe that had befallen me. Pat had a bite to eat. The old bush rat declined our offering.

"No, thanks." He said, "I've lots to eat. The nearest store is only 30 miles away." After that he disappeared.

Said Pat, "We might as well be on our horse again. Have to stop for gas somewhere. Can't make Port Arthur with what's left in our tank. That damned side wind is holding us up!"

We swung the plane round, and giving his motor the gun from the start, somehow Pat squeezed through that narrow inlet and into the air before the whitecaps hit our floats.

Rossport was our next stop. Like the last landing place, it was in a bay off the big lake. There was a pier with some boats tied to it, but enough room was left for our plane. The whole population of that forgotten-looking spot came down to see the unexpected visitors.

"Could we have some gas? Naphtha if no aviation spirits."

A husky fellow, apparently the big shot of the town peeled himself out of the crowd.

"Sure, we'll give you some naphtha, but we haven't got a pump or hose long enough." After a little while a truck with some barrels on it backed down to the shore, but not onto the rickety dock. The good people of Rossport formed a bucket-brigade and all went smoothly. Thank you, Rossport! With your help we got to Port Arthur all right, where again we refueled our plane, and refreshed ourselves with some nice cool milk.

It was Pat's plan to make Kenora by nightfall, taking the CPR track as his guide. But somehow, with the compass not working properly, there was a big air bubble in the contraption, we must have gotten onto the wrong track. Finally, after what seemed a too-long flight, we saw a lake dotted with islands. In two places seaplanes were tied up. That must be Kenora.

We landed and taxied towards a pier. A fellow came running while I tied down the plane.

Pat calls from the cockpit "Is this Kenora?"

"Hell, no! Sioux Lookout."

"Holy cats! How did we ever get here?"

"You tell me, you are the pilot."

"Well, no use worrying about that now. We got here. Good day's work too. Thirteen hours in the air."

A taxi driver, bootlegger, and boarding house owner took us into his loving care. We had a bath, some beer, something to eat and a room for sleep. Oh, yes, some company could be arranged if we so desired!

We didn't even know how to pay our considerate friend. All we had were travelers' cheques and nobody seemed to consider them good money. After some difficulties, the bank manager was located, and he, perhaps a little grudgingly, after all it was Sunday evening, solved our troubles. Bright and early the next morning we were ready for the next flight, with Winnipeg our goal.

Not the slightest breeze was stirring. Not a ripple disturbed the glassy surface of the lake. We have to try it just the same. Pat gave her the gun, rocked his plane desperately to get it riding on the step. After many tries, the nose stayed up.

"Hold on. Here goes nothing!" came Pat's voice from the speaking tube, as he pushed the throttle forward.

A wooded island came rushing towards us it seemed. Were we going to make it? Yes, we were off. Something black and shiny to our right. By fractions of a second and a few inches, our right pontoon missed a half-submerged floating log, and the treetops on the island seemed awfully close. A narrow shave, but we climbed, climbed, and headed westward again.

It's a long dry hop to Winnipeg. Pat told me we would be flying high. In the cabin it got very chilly, and I was glad when finally, the Gateway to the West came in sight, and we landed between the muddy banks of the Red River near the Canadian Airways shops. Pat got a mechanic to check the controls and the motor. In town we met some friends and had a few beers. Winnipeg was still the same place as in 1930 when I first saw it as a green immigrant, but somehow it looked different now that I have found my bearings in Canada, and the future looked promising once more. I remember riding the rails to find work on the way out west and pulling stumps on a farm near Whitemouth in 1930.

The reunion between Pat and his old chums was going full blast. Western hospitality found its expression in more and more rounds of beer till the table wouldn't hold any more. Early in the evening I stole away from the friendly meeting, headed for my night's quarters in our plane. Pat's voice roused me from deep, dreamless slumber while the mist was still rising from the lazily flowing Red. Not a breath of wind. There was to be a test flight after the readjustment of the control cables to correct the nose-heaviness. The take-off would be difficult. Anxiously we were watching when Pat disappeared around a bend, throttle wide open, and finally came skimming over the far bank, ruffling the hair of some early golfers.

"I just made it by the skin of my teeth" he reported, "but I couldn't take off with you on board." After a little palaver it was decided that I take some of our load and myself by bus to Lac du Bonnet, about 60 miles east of Winnipeg.

The road was familiar. In 1930 some chums and I rode it going to Slave Falls to try to get a job on the power project then under construction. Many memories were called to life again as the wheezy old bus rumbled along.

At Lac du Bonnet I found Pat waiting for me to join him for lunch. We didn't waste much time. It was Tuesday noon. By Thursday we were expected to reach Edmonton. A long way to go yet, following the water route with our seaplane. Our course will be west of north across Lake Manitoba, then along the west shore of Lake Winnipeg towards Grand Rapids, where we intended to refuel. When we finally approached the marshy and sandy wastes of the delta of the Saskatchewan River, we saw to the north a clear bluish-white line. We had caught up with Old Man Winter.

"No chance to land. We have to try to make The Pas," Pat told me.

For hours, it seemed, we were flying over the delta's bewildering maze of little streams. There must have been thousands of them, lined by gray clay or yellow sand, encircling greenish blobs of grass, reeds, and willows, until at last we were above the river proper, where it looks more like "swift flowing water."

The sun was still high in the sky when we reach The Pas. Like a picture of bygone days an old sternwheeler went gliding past the dock where we were refueling. Pat looked at his map, scanned the skies, "OK. Let's be on our horse again, we can make Prince Albert while there's daylight."

The Saskatchewan led us to our goal for the day. Twilight was filling the river valley when we pulled our faithful plane, tail end first, onto the muddy shore just below the town, which seemed to be the lovers' lane of Prince Albert.

Towards noon the following day we were circling for a landing at Cooking Lake seaplane base. We met Pat's friend George who took us to the MacDonald Hotel in his car. Pat and I took a large room together, and this in a short while turned into a very busy and crowded meeting place. Friends, people from the North, bush-pilots, and their friends all flocked together and had a few ice-breaking drinks with us. Rye, lime juice and ice were mixed in larger and larger quantities to meet the demand. We told our news and pumped the assembled experts on the North for information. Shopping lists for the thousand and one things we would need during the summer were compiled. The bulk of the supplies would go by rail and barge to Beaverlodge, which would be our home base, as soon as the water route was opened after the ice melted but we needed to have enough provisions flown in to tide us over.

Our airplane mechanic, "Out our way Reg" as Pat introduced him, showed up. Pat set him to work immediately itemizing things needed for the maintenance of the all-important plane. Several days were spent rushing around from the Hudson Bay Company to Marshall-Wells, Ashdown, Uncle Ben and so on, but when, at long last, we thought we were all set and ready, the weather turned against us. Ceiling zero and rain, rain, rain. Rye and lime juice consumption jumped while we sat it out.

In the meantime, we decided that I would go by train to Fort McMurray. That way I could take some of our paraphernalia with me without cost, and we would save one round trip. So, I bought myself a ticket for a memorable train ride on the Northern Alberta Railway. I didn't cherish the idea very much, but looking back now, I know would have missed a great experience; a trip on the *Muskeg Limited* as Northerners call a trip through the swamp.

CHAPTER III
TO THE END OF THE STEEL

A BIG CROWD was assembled at the CN Station the morning of my departure. Some wearing city clothes, others in Mackinaws and breeches, windbreakers, and overalls. Young and old, tall, and short, a colorful assortment of humanity, not to forget some French-Canadian Sisters and black-bearded Priests, all headed north.

It was quite an impressive train. A few old-fashioned Colonists' Coaches dangled at the end of a long string of freight cars. "All aboard!" shouted the conductor. Farewell scenes broke off and the passengers, all hugging parcels, valises, and duffel-bags, climbed onto the train. I looked around for a seat near a window.

"Better sit down," somebody called to me. It must have been an old-timer. From way ahead the crashing of couplings rolled towards us, swelling to an earsplitting crescendo. Before I could find out who gave me the friendly warning, our coach went berserk. The train was shuddering and jolting the passengers. I was thrown into the next seat. Some were less fortunate and landed amid the tumbling baggage in the aisle. The train's convulsions lessened, but hardly have we gained a little speed when once more the crashing noise approaches. We got a good shaking down and the train stopped.

"This stop is Dunvegan Yard," said my neighbor.

"What are we stopping for?" I asked innocently.

"Don't know. Say, is this your first trip on the Muskeg Ltd.?"

"Yes, it is and I already hope the last one."

"Well, you are in for some surprises. They usually spend about an hour here. They stop more often than they go, that's why it takes them 22 hours to cover some 300 miles."

The full truth of that statement was soon proven. The train either stopped or proceeded at a most leisurely pace. Each slowing down or speeding up was accompanied by that infernal racket. The rear and front ends never seemed to agree about stop and go.

At lunch time this led to a few minor catastrophes. Even our spoons got rattled on their familiar path from plate to mouth. Nobody ate with his knife, I am sure. Poor Blackie, as the porter was called by the regulars, who served the meals had an awful time balancing his trays. One moment I contemplate whether it is worthwhile to dissect the second half of my tough steak, the next moment it has jumped into my lap. Farther down the aisle a feminine voice cries out. Blackie comes running. One of the good Sisters had an accident. There had been another terrific jolt, and the soup had left the plate. Blackie was full of apologies for that nasty train, promised a new bowl of soup, but who wanted more soup?

I took a chance on a piece of pie offered by Blackie. Then when I had won my battle with the elusive dessert, I fell into a conversation with my neighbor on the other side of the aisle. He was a tall, husky looking fellow with a friendly grin, and a slight Scandinavian colouring in his voice.

"Where are you going?" opened the talk, and from then on, the hours were flying. There was a fellow who knew the North and had the gift of talking about it to an ardent pupil.

"I am going to Fort Norman," he said. "I run the show there for Imperial Oil. You know, they have an oil well down there and a small refinery. I suppose you will go after gold mainly, but believe me, there is more than gold in the North, and little has been done about it so far. There is oil at Fort Norman, just oozing out of the ground. I have covered a lot of territory every summer and found many a spot where there are unmistakable signs of it.

"We have just one well, and during the summer, refine enough for the Eldorado Radium Mines. Someday the country will need it, then they will find out what's all hidden down there. You know, they don't even have to go that far. Ever hear about the Tar Sands at Fort McMurray?

"There's coal up the Liard River, there is waterpower in abundance. I tell you the surface hasn't been scratched yet. Talk about the North country? Brother, there is an empire to be conquered."

While I was listening to this fascinating tale about one of the greatest undeveloped areas in the world, of which even the Canadians on the whole knew little, I had noticed a young chap in his early 20s, a few seats away from us, looking forlornly out of the window at the dreary landscape studded with second growth bush, mainly poplars. My new friend, following my glance remarked, "Must be his first trip too. He looks damn lonely. Maybe he thinks this is the North."

"I'll ask him to join us. That might cheer him up." I went over. The poor boy had the blues alright, and he joined us only too gladly.

"You know," he told us, and once he started it poured forth in a torrent, "my old man is sending me North. I am to go on a prospecting trip to Lake Athabasca with Colonel X. The old man thinks I am too soft, and a summer in the bush will toughen me up. So here I am, and by what I've seen so far, this surely is one hell of a country, and I am damned homesick. Ah, well, it's nice of you fellows to ask me over, though. What are you talking about?"

"The North, of course," I replied. "Here is an expert. Just listen!"

Soon our little man was just as eager as I, and it wasn't much later that we were surprised, by "Last call for dinner."

"Never mind that," said our friend, "We'll eat at Lac la Biche, we ought to be there shortly. We'll get us a darn good feed of fresh pickerel. Nothing like it."

Just then with terrific jolting and rattling the train came to a stop once more. "Well, of all the confounded so-and-sos," exploded our Fort Norman friend: "Here they pull the same trick again. There goes our nice feed of pickerel. They want us to buy our meal on the train!" And so it looked. There wasn't a station in sight anywhere, not even a section gang's shed.

Someone said: "Well, here goes our engineer again to look at his traplines." That explanation had previously been given for the innumerable stops right in the middle of the wilderness, but some said, it was just Mother Nature calling him. Whatever the reason, it turned out to be a long stop, so long in fact that we finally gave in and answered the last of the last calls for supper.

"At least they let us eat in peace," somebody daringly said after the soup. A rash statement. The train's whistle blew, couplings clanked and crashed, the Muskeg Ltd. was on its jerky way again. A few miles of perturbed motion, then whistle, crash, slam, bang. Another stop. Well, what do you know? Lac la Biche! The steak had been dry, but the comments were juicy.

Anyway, it was a relief to get onto terra firma, to stretch our travel-weary limbs. Towards the lake we made our way. Our friend from Fort Norman pointed out the hotel where we were to have enjoyed a real treat. Two boys walked by with a nice string of fish: "See what I mean, pickerel!" After walking around for a while, we sat on the dock where some natives were busy catching their supper, until finally the whistle called us back to old "slow and noisy." A night to be spent on it was not an enchanting prospect, but somehow even that came to an end and early in the forenoon we finally chugged into Fort McMurray. End of steel. Threshold of the real North.

Never before was I so glad to get off a train, although it meant parting with my newly found friends.

"So long and good luck."

"Look me up at Fort Norman if you get the chance."

"Meet you around Lake Athabasca!"

Pat was standing near the station building, leaning slightly into the north wind, a wide grin welcoming me off the train.

"So, you made it! Got everything?"

"Yea. Some train, some ride!"

"OK Let's get going. I got a taxi here, best goddam taxi driver I ever met up with."

"Hey, Red. Meet a friend of mine, picked him up in a beer parlor," Pat said introducing me.

Red, a jolly buxom lady with flashing, laughing eyes got us to Fort McMurray in no time over a smooth tar/sand road, while Pat was holding forth: "Here is the plan. As soon as I can get away, I am going to take Dick into Beaverlodge. B&M has a camp there. Dick is an old prospector the boss wished on us. I don't think he likes me, but he likes the beer I've been buying for him, so we won't worry. I'll take you second trip in with some more of our equipment. After that, I'll take Reg and the rest of the stuff. I'll be back tomorrow, noon at the latest. Get all our stuff down to the Mackenzie Air Service shack in the meantime."

In the beer parlor of the Franklin Hotel was the meeting place for people heading north. It was a busy place that morning, crowded with prospectors, trappers, bush pilots and inhabitants of Fort McMurray, we met up with our future companion in the bush: Dick, greeted neither of us with great enthusiasm, but after a few more beers, he melted somewhat, and grabbing me by the shoulder whispered very confidentially into my ear, "What do you think of the theory of evolution?"

Nonplussed, I replied, "Who am I to argue with Darwin?"

"By golly," he said, "that's the name of the guy in that book, couldn't think of it right now. We'll talk about that when we are in the bush, gets damned lonely at times."

After lunch we all went down to the sea plane base on the Snye River where Reg had our plane all gassed up and loaded, with just enough room left for old Dick to crawl in. Pat climbed into the cockpit, pulled down his goggles, Reg on the float, right hand on the cowl, left on propeller tip sings out, "Contact?"

"Contact."

The motor hammered out the opening cadenzas, then settled down to a deep-throated roar. Mighty symphony of the awakening North.

"OK Shove her off, keep your nose dry in the meantime."

Towards the Clear Water River Pat taxied his ship, turned into the wind and gave her the gun. He got her on the step, and he was off.

CHAPTER IV
NORTHWARD

NEXT DAY IT was my turn. Once more the cabin was filled with grub, tents, tools and sleeping bags. There wasn't much wind. After several, vain attempts to get the heavily laden plane off the sluggish waters of the Snye River, Pat turned into the Clearwater River and got us airborne just before we got onto the Athabasca itself. Banking steeply, we turned, climbed, and then leveled off, and with the watery Highway to the North meandering below, headed northward, ever northward, the small shadow of our thunderbird racing ahead of us. Sunlight was painting silvery filigree on the gray-green water. White sandbars here and there, thick bush along both shores, broken by muskegs and sandy patches further inland, and little frustrated creeks and rivers futilely turning here and there, ending up in sloughs and potholes, made up the scenery. No living being could be seen from above, it was like a forgotten world, under its uninviting exterior hiding, who knows, perhaps the greatest treasures man had ever dreamt of.

Alone in the cabin, with my thoughts crowded with a multitude of new impressions of the immensity of this Canada. I kept on gazing, with a strange fascination, at the seemingly endless scenery below, forever changing, but always the same. When would man conquer all this, following the trails a few Indigenous people and trappers had blazed, to fill the immense vacuum that stretched for thousands of miles in all directions? Now we were a part of the first few thousands who had been attracted by the North since the first white man had come to this continent. I felt for the first time a sense of adventure, a feeling of happiness, and expectancy, and I felt grateful to a fate that had led me across the path of the fellow who sat behind the controls of our plane now, whose question, "You want to come North with me?" had brought me here.

This was the overture to a new life. How would it turn out? I didn't care. This was it. I never felt so fully alive before, this must be what I came to Canada for. Forgotten the Depression years, spent in a sleepy Laurentian town. Forgotten the endless attempts to land a good job. Here was a job for men, to hew homes out of wilderness, to blaze new trails to a better future. It was like a prayer in my heart, a prayer that we might be strong enough for the task, and a prayer of thanksgiving too.

My reverie was suddenly interrupted. Pat's voice came out of the speaking tube, "All, OK? I'm heading slightly east now to pick up the mouth of the Williams River. From there we'll cross the Big Lake to Cracking Stone Point and Beaverlodge."

Flatter and sandier becomes the scenery below, the meandering Williams River came into sight, then a whitish gleam on the horizon: Lake Athabasca. Once more we flew over the delta of a river, where it split into a thousand small branches, each trying to push its own feeble way toward the big water.

Massive floes of ice were still dotting the waters of the big lake. Pat took us higher up. On floats it would be a risky business to go down, too much ice, and the water was rough. White spray flew where the waves broke against the edge of the ice. The far shore presented a new picture even from 40-50 miles off; rocky hills serrated the horizon. Not long and we circled over the few scattered shacks and tents of Beaverlodge, later incorporated into the Goldfields mining area, then glided towards the dock of the B&M camp, where we tied up our faithful plane.

Pat and I walked towards the camp, "Wonder where the old bush rat. D.G. got to?" says Pat, "he must have heard us coming." Just then from a group of tents at the end of the trail, came the sound of ringing steel, and then we saw the cook in his white apron banging away at a triangle of drill steel hanging outside the biggest tent, "Come and get it!"

"Just in time, Hermann, we'll get us something to eat, then unload the plane and I'll be on my horse again!"

The cook saw us coming, squinted his eyes, then, "Pat, for the love of Pete, what in hell brings you here?"

"Hi, Charlie! What are you, old reprobate, doing here? How is the lemon extract?" came back from Pat.

"Never touch the stuff, Pat, honest, that is, not since the last time."

"Got anything to eat for us?"

"Sure, sure, come in, the staff is eating now, some of your old friends."

"Friends??" Pat answered, with a big question mark. The welcome seemed a little chilly.

"See what I mean, Hermann?"

"No, I don't, quite."

"Well, if you had read my diary at the hotel, as I told you to, you would," Pat said to me.

Pat's jaw was sticking out somewhat more than usual, and there was a sort of "Want to make something of it?" look in his eyes.

That didn't spoil our appetite though, and Charlie's grub tasted good. Old Dick showed up after the pie, gave us a hand with the unloading of the plane, and Pat headed south again, after spitting out a string of cryptic but powerful descriptions of a certain outfit and certain so-and-sos. Old Dick seemed to know more about it, but if he were waiting for me to ask him, well, he just could keep on waiting. Pat was the guy who had asked me to come North, he had trusted me, and I had liked him from our first meeting.

He was a fighter, and fighters are needed when opening new frontiers. You don't ask questions then; no time to think about the past, the future promised to be interesting. The future, prospecting for gold! I thought maybe I'd better start right then to get my bearings. I had a little knowledge of minerals, rocks and formations from my years at the agricultural college, even dimly remembered one of the favorite subjects of my old geology professor, The Laurentian Shield, oldest mountains in the world, Precambrian range, etc. But how do you look for the elusive Gold?? While these thoughts went through my mind, we had reached the spot where Dick had put up his one-man tent on an exposed spot near the shore, west of the camp.

"Pretty breezy here," I said.

"Sure, keeps the mosquitoes away."

My first lesson in northern survival. I would be a willing pupil if the old fellow wanted to teach me. But old Dick wasn't thinking of the future, his mind was roaming around in the dim past.

"Now about this here evolution," he came back, and that ended my hopes of finding out about prospecting, that sunny spring afternoon on the shores of Lake Athabasca. Old Dick surely had an inquisitive mind, and I fully regretted not having paid better attention to my lectures in days gone by. Darwin, forgive me, if I invented a few theories of my own, there was no getting out of it. We were still at it, when we faintly heard the cook calling for supper, and it was the last thing before falling asleep in my pup tent with the wind and the waves singing us a lullaby.

Toward noon the following day Pat came roaring back, side slipped into the bay, taxied toward the dock, climbed out of the cockpit, and threw us a rope. We pulled in the plane, wondering what was up.

"Where is your load? Where is Reg?" I asked

"Dropped him at Beaverlodge Lake. Found the best goddam camp site, sandy beach, and all. I'll take you fellows over now. Get your stuff aboard. I'll get the rest later.

CHAPTER V
FIRST CAMP

A SHORT FLIGHT and the floats ground against the sandy beach of our first campsite. It was a beautiful spot indeed. To the north, stretched one of the most picturesque lakes I had ever seen, sparkling clear water, dotted with a few wooded islands. To the south stretched a sandy plateau with jackpines, a rocky range rose abruptly to the west, and looking east one saw the highest elevation, Beaverlodge Mountain, rising like the sharp prow of a liner. But there wasn't time to take in the full beauty of our surroundings.

Work was calling. Clear the ground. Put up the tents. Build a cache high off the ground for the grub. Cut firewood. Install a latrine at a respectable distance. Our new life had begun. Life in the bush.

By the time the camp was established the sun was lowering in the west. Old Dick had built a fireplace near the shore and brewed us some tea. Hardtack and bully-beef made up the spartan menu. But we should soon do better. Dick mumbled something about sourdough and bannock, "bennick" as he called it.

Then the mosquitoes launched their first attack, and big vicious fellows they were. We all took refuge behind the mosquito net of Pat's tent, there to talk, smoke and to discuss our further plans.

"Best thing," Pat suggested "is to stake some claims to hold this campsite. It's a good spot. Here, look at this map. See this portage from Lake Athabasca to this one? We are not far from it. We'll have our supplies unloaded in this bay if they can push the Hudson Bay scow in that far. Won't be anything to it to get our stuff across the portage to camp if that works out. You fellows have any better ideas?"

It looked all right to us. There was no comment from Dick. We soon got the impression anyway, that the old sourdough didn't seem to hold a very high opinion of us greenhorns. There had been a few caustic remarks about the equipment we had brought. Worst of all, we had brought no yeast cakes.

"Baking powder won't do?" I asked.

"Cripes no! Man, that's pizen! Makes stones in your stomach!" Dick exclaimed. Pizen being poison in Dick's speech.

We didn't worry though, as long as we could get a few pointers from him as to how you find a gold mine, if he knew.

A light breeze, sighing through the treetops, barely ruffled the surface of the lake, now seemingly turned into a cauldron of molten gold by the westering sun, with the trees on the far shores etched as with steel pens against the flaming horizon. Silence fell over our little group, only to be broken by the war chant of thousands of mosquitoes, and the hum in our ears; not yet accustomed to the otherwise quiet of the North.

"It gets you," they say down North, not trying to explain, just stating it as an undeniable reality.

Maybe it was then that began the infiltration into our souls of the undefinable something, that would never again be lost, the something that makes a man return to the North, in spite of hardship, in spite of loneliness, again and again, even if it can only be in his thoughts. Wherever fate may lead him, the sound of an airplane overhead, of an axe biting into wood, waves lapping against the shore of a lake, unfailingly leads his thoughts back to those immense reaches of the lake-dotted wilderness where only a few so far have built their camps and broken rock. Whoever has tried it, he will accept it, and all his life there will be a yearning in his soul to make camp once more on the shores of the crystal-clear lakes of the never-forgotten North.

The long twilight of the North was still lingering when we finally crawled into our bedrolls, Pat and I sharing the netted tent. I had a feeling Pat had something on his mind.

When we were alone, it came out, "What do you think of the country?" asked Pat.

"Looks good to me. I never thought, hearing about the North, it would be so beautiful. We sure got a swell campsite. As far as I am concerned it's the biggest thing that has happened to me in Canada. Thanks for asking me to come."

"That's OK. But what I want to know from you is something else. Do you think this is a place for a woman?"

"What????"

"I didn't think that would be a surprise to you. I mean to say you've met Maud. I am going to send her a wire tomorrow, to meet me at Fort McMurray, say a week from now."

"Well, I am sure your mind is made up about that, so whatever I may say has little influence on that question. You didn't give me much time to think about this angle, but offhand I'd say there is no good reason why Maud shouldn't spend the summer with us here in the bush. I'm sure she can take it. We'll just have to see that we make this camp as comfortable as possible."

"Well,," Pat replied, "I am damn glad to hear you say so. Old Dick will be a different story. I am going to tell him in the morning. Now we had better catch a little shut-eye."

Dick's face was quite a study next morning. Not that he said much. Just turned back to his fire, where he was "biling" some tea, muttering under his breath.

"Wimmen in camp, that's pizen. Pizen I tell ye."

The old boy surely had his strong opinions, but others weren't lacking them either in their make-up, as became evident very soon. Pat's jaw seemed to be sticking out a little more than usual that first morning at our first camp.

Irish eyes weren't smiling, when with a hint of the old brogue in his voice, he delivered his pep talk, "Might as well tell you right from the start. I am going to run this show. If there's anything about that you guys don't like, you'd better speak up. So, we all know where we are. I don't want any goddam yes-men in my camp. Of those bastards I've seen plenty in my short life. And that's the longest speech I've made in a long time, and we'd better get busy."

Before the day was much older, we were too busy for further speeches. A dock had to be built far enough into the lake, from the end of which we could get clear water for our cooking and washing, and service the plane too.

Pat flew off to Beaverlodge together with his "grease monkey."

Dick then proceeded to initiate me into the intricacies of staking claims, salting his explanations with candid remarks about greenhorns. However, once in a while he popped another question about evolution at me. So, I proposed a bargain, "You tell me all you know about prospecting and I'll tell you all about old Darwin and the arguments he started," hoping silently that where memory might fail, imagination would do the trick.

Blazing our westernmost claim line on our way back to camp, late that afternoon we got onto higher ground and rocky outcroppings. Dick apparently didn't like the look of them.

"Poafry, nothing but goddam barren poafry. Now you take Northern Ontario, there is a place to look for metal." Porphyry is igneous rock with large crystals and not good for mining. He meant porphyry, but it always sounded like "poafry".

"OK Dick. But we are now in northern Saskatchewan, and somebody found gold not far from here. What signs do you look for?"

"Well, what I like to see is greenstone dikes with quartz veins cutting the formation and then, you look for base mineralization along the contacts."

I thought, *what an answer?!*

Things weren't getting any clearer to me in a hurry. "Are there any books on this, Dick?"

"Books, books! Some guys think they can learn everything out of books. Seen many a guy in the bush who thought he knew all there was to know. Sure, guys from a University, throwing around with fancy names. I'll tell you; you never find a mine that way."

"OK, how do you tell?" I asked.

"Well, a prospector, a real prospector. I mean, he just gets a feeling, the country looks right to him. Now you take Northern Ontario…"

By and by we got back to camp, which was in Northern Saskatchewan much as Dick would have liked to be in Northern Ontario. I had learned about staking a claim, but the finding of the gold, that was a greater mystery than ever. Then a roar from the south and Pat came skimming over the treetops, side-slipped, straightened out a few feet above the water, put the plane down in the smoothest manner, and taxied toward camp. Another full day was almost over by the time we had our chow, and the mosquitoes drove us once more behind the fly bars of our tents. Smoking contentedly, we looked out over our lake. Little waves were rolling lazily onto the sand, mosquitoes were humming, Lullaby of the North! Peace blessed Peace.

Suddenly a fierce unearthly cry, piercing the quietude, tore us out of our reveries.

"Ah, hell, just a loon," someone said. The first wild eerie cry will always catch you unaware, sounding like the last desperate outcry of a lost soul.

Another day was dawning soon, and we had a full program ready. The four of us finished blazing the claim lines first, then Dick and Reg went exploring while Pat and I got busy building a bed out of peeled posts.

"Might as well be comfortable," said Pat, as he dug out of his duffel bag an inner tube he had bought in Edmonton. I didn't want to seem too ignorant, therefore hadn't asked any questions about making the bed comfortable before.

"This will make damn good springs, I hope," he declared, and proceeded to cut segments out of the tube. These were attached to the frame, rope lashed through them from side to side and crossways, and there was the bed.

"Floating power, or something, what?" I turned to Pat when our handiwork was finished.

"What do you think of it, Hermann?"

"Darn good idea, and compliments to your foresight but will it work?"

Pat lowered himself gently onto the connubial contraption, a smile of utter contentment stole over his mug.

"Perfect. Just perfect! Try it yourself! Perfect!"

Gently, I stretched out, too. It sure was a vast improvement over lying in your bedroll on some spruce boughs, but some doubt crept in.

"Remember the cartoon, I think it was in Esquire. A little fat guy and a little fat lady sitting on the edge of a bed in a showroom, with innocent faces looking up to a towering salesman. Says the little fat guy, "This, of course, tells us nothing!" I looked at Pat.

"Holy Moses, but you can shatter the sweetest dreams. This is no place for pessimists."

"Come, come! I dare say I was an optimist in this case! I figured on two on this here couch!"

"Have it your way! If this thing breaks down, YOU'LL have to fix it," said Pat.

"OK What next? A tent and a bed we have. We'd better fix up something like a cook shed or kitchen. We guys can manage with Dick's hole in the ground and some forked sticks, but since we are expecting a lady for company…

"Tomorrow I may have word from Maud. Maybe Alex, that's the trader at Beaverlodge by the way, a square fellow, may have a tin stove and a tent we can buy. If not, I'll get them at Fort McMurray."

Time went by quickly, and soon the day came when Pat went sailing south again to meet his chosen one. Old Dick got grouchier than ever. Even steering the talk to evolution didn't brighten his moods anymore.

But the sun was shining, and a whole new world was waiting to be conquered. No time to worry about Dick's idiosyncrasies. A few days later, the drone of an approaching plane heralded Pat's return. Westering sun was tinting the red and blue of our plane with gold, making little rainbows out of the propeller-caught spray as Pat raced toward camp on the step. The nose came down, a few more turns of the prop, and silently the thunderbird drifted onto the sandy beach.

Pat, a puckish grin on his face, goggles pushed up to his forehead, peeled himself out of the cockpit, "Don't you guys ever shave?"

Out of the cabin came the newest member of our party, raven-haired Maud. And so, we were five. Joyous greetings! Even Dick tried to crack a smile, more from embarrassment than enthusiasm though. Gloom soon crept over his craggy features.

"What a beautiful spot!" exclaimed Maud, who was proudly presented and guided about the camp by a very happy Pat. A few bottles of beer appeared out of Pat's bedroll.

Bannock à la Dick, liberally sprinkled with wood ashes, fried dried spuds and bully beef seemed like a festive meal.

"Where is the stove, Pat?"

"No room in the plane. Alex has a small one. I'll pick it up some day. The Hudson Bay scow will be here in ten days or so. I ordered a lot more stuff by wire, a big stove too, and a tent 9 x 12. Find any gold yet?"

"I thought Dick was the guy to do that," I said.

"Hell, no! You'd better get busy. It's up to you."

"Me?"

"Sure, beginners' luck, you know. How'd you get along with Dick?"

"No trouble. He isn't happy though. Doesn't like the look of the country, he says."

"We'll soon cure the old so and so. Tomorrow I'll take him along. Do some prospecting from the air. Find a likely-looking spot where he can start cracking rock."

"Say, Pat, what does the rock look like where the first gold was found, over at B&M?"

"We will go over some time and do a little investigating."

Northern sunset magic was weaving its spell again. The talk died down. Before our eyes once more the lake turned into a shimmering expanse of liquid gold framed by the wooded shorelines etched in deep purple and black. Only the crest of Beaverlodge Mountain was still illuminated by the last rays of the setting sun, then it, too, turned into a purplish silhouette. Night had fallen but darkness didn't come. A luminous sky remained, light greenish blue with an orange-yellow tinge along the horizon. Summer night of the North.

"What do you think of it, Maud?" Pat broke finally the silence.

"I cannot say it in words. It is so beautiful and so different from anything I expected. The North, whenever I heard that before, I imagined something barren and desolate. Now I know."

"How about you, Hermann, not sorry you came?" asked Pat.

"Sorry? After seeing this golden lake! If that's all I ever find, I shall be content."

"You are a dreamer. Don't forget our bargain to find me that mine. This is a cruel commercial world. Don't forget it." That was Pat again.

"I will try my best, but as for that dreaming stuff, didn't you tell me your favorite melody was *Liebestraum*? Good night and sweet dreams." I turned off the flashlight.

Many exploratory flights were undertaken by Pat and Dick during the following days, while Reg and I made further improvements around the camp, cut a trail to the portage, and tried to build a mosquito-proof privy from poles, bags and packing cases. We used great care, by now the worst predictions of friend Charlie in Toronto had come true. Pine tar and Citronella was all right on your face and neck, but when Mother Nature called …

One afternoon Pat and Dick stayed away longer than usual. When they finally returned Dick seemed to be in a better than customary mood.

"We found some promising rock not far from here on Fredette Lake. I am going to take Dick over there again to have a good, long look around."

Dick dug some rock samples out of his pocket and quite proudly showed them to us. He explained: "See here, blue-green stains, and yellowish mineral, that's copper pyrite and this here, looks like silver, that's galena. Good indications."

On the way back they had stopped at Alex McIvor's store and picked up a small two-hole tin stove. Just in time, because that same evening a short but violent storm blew up, and our outdoor fireplace proved to be a poor kitchen on a rainy night.

Life in the bush taught us new lessons every day. This was an important one. Others followed and by and by we learned to make life in the bush as comfortable as possible, in any weather, at any place. Our tin stove became a symbol of that endeavour.

At Alex's, Pat had picked up some more information about our district.

"A trapper, Swede or something, has a cabin at the north end of this lake. We'll drop in on him some time. A lot more prospecting parties are coming in. Most of them have put up their tents near the B&M One camp I noticed right on the portage. We'll probably have some visitors soon.

CHAPTER VI
FIRST STRIKE

NEXT DAY WE set up Dick with a camp at Fredette Lake so he could continue prospecting.

On our way over we took a good look at the B&M show, from above, where diamond drilling was under way.

"Look for breaks in the formation and quartz veins," was Pat's advice. At low level, in wide circles, we swept over Tazin Lake, towards Cracking Stone Point covering more and more ground.

From then on Pat and I did the prospecting from the air.

One day we were flying in a north-easterly direction over a section where a forest fire must have swept through some time ago. Most of the rock was exposed. Suddenly I glimpsed a pronounced rusty streak running with the formation alongside a narrow lake. I grabbed the speaking tube and said, "Hey, Pat, see that down below?"

Before I had finished, Pat had kicked the rudder and pushed the stick. "Hold on, we are going to land. Looks interesting."

Down we went, drifted toward the shore, tied up our plane in a hurry, and prospectors' picks in hand scrambled toward that rust-coloured streak. Reddish brown rust was all we saw on the surface of the thoroughly disintegrated vein. Feverishly we picked into the crumbling mass.

"There must be lots of mineral here," Pat remarked as we tried to get deeper, to more solid rock.

"Here, look at this!" Pat called out, "Copper pyrite, iron pyrite, even specks of galena!

I bet you anything, this stuff will be good."

A less rotten lump of rock, cracked open, gave us the answer.

"Look at this! Look at this! Boy, oh boy! This sure looks rich." Pat was excited.

The more we dug in the more mineral we found.

"Just like a damned jewelry store. This takes some looking into. Let's see how far this goes. Must be 15 feet wide, but how long is this thing? It's running pretty well north and south."

To the south a bay of the next lake cut off the lead, so we turned north picking up the same oxidized streak again along the edge of a trench-like break. The going wasn't easy. A maze of half-charred fallen trees filled the gully, little clouds of rusty and black dust rose under our feet, we got blacker and blacker, but who cared. Skirting small bays, where our lead disappeared into the water, we picked it up again on the opposite shore. After we had travelled quite a distance the overburden (soil, grass and leaves on top of the rock) became heavier, then the lake widened out.

"That's as far as we can go today. Let's get back to the plane, take some samples and make for home. We'll be back tomorrow. We need more men with licenses and get our own licenses and those of our proxies. Later on, we need more picks, shovels and then drill steel, sledges, and dynamite. God bless my teeth, but we'll be busy."

So back to camp, Pat wagging his wings as we buzzed our tents, bringing the rest of our little gang to the shore to welcome us. "Hi there!" hollered Pat. "We have struck it!"

I threw the rope to Reg, who hauled us in, then dove back into the cabin to bring out our samples. Pat dumped them onto the sand. "Have a good look at that! That's pay ore or I'll eat my shirt!"

Maud and Reg looked at our treasure.

"Is this gold, Pat?" asked Maud.

"No, but it's ore, and there is a big body of it. What do you say, Hermann?"

"Well, it looks good to me, but I don't know much about this business yet. If Pat says it's good and he has done prospecting before, that's good enough for me."

"I got damned curious about what is down below, I must say."

We were getting down to serious business. From that day on, there was urgency about all our actions. So far it had been something new, exciting, filling every waking hour of our life in the northern bush, like exhilarating sport. Now: Let's get to work.

Pat's explanations soon made it clear to all of us: "First thing: we got to protect ourselves by staking claims. If we don't and other parties hear our shots they will come flocking around, and maybe stake the best ground right under our noses. So, we will run the outside lines first, get our No. 1 posts in and run the other lines later. We have 60 days to record our claims. By then we'll know what we've got."

"Hey, Reg!" to the mechanic, Pat called.

"Yep."

"The plane ready?"

"Ready as she ever will be."

"OK Let's be on our horse. Keep your noses dry!"

"Contact!"

"Contact!"

Pat was off to Beaverlodge, soon returning with another new member: Stuart, the former Mountie, a broad-shouldered gentleman who made himself at home without delay. A solid-fisted handshake, a square look into each other's eyes. No fuss or formality. A guy throws his packsack and bedroll down, and there he is.

"I've lined up two more fellows," said Pat, "might as well get them now from Beaverlodge. Better get your stuff ready. First thing in the morning I'll take you all over to the show." The show is the place where indications are good that we might find gold. Axes, picks, shovels, crowbars, prospectors' picks, sample bags, some grub, a pail, and a pot. No sooner had we lined up our paraphernalia, when Pat arrived with his next load; soft-spoken, broad-shouldered, blond Ronny and dry-humored, dark, wiry Art.

"There's your new crew, Hermann, pretty good-looking bunch of renegades. Now you guys understand what it is all about. You will get good pay and grub. There will be plenty of work. We have to make a mine this summer. Our show looks good, we'll have to prove it up, or find another one. We'll give you a fair break, I've always believed in taking my friends along. If we make good, and there'd better be no 'if' about it, the sky's the limit. I'll bet you before long we'll have the best goddam mining camp in this country. What's going on in our camp is nobody else's damned business. If there's anything you don't like about the way this show is run, better spit it out. I won't have any yellow-livered yes-men in this outfit."

By the look on the faces of the newcomers this was a pretty overwhelming speech, and maybe our new comrades didn't quite know what they had let themselves in for, but outspoken Pat could turn on the old wonder-working charm too, and he was at his best when we were all sitting around our campfire.

The chill of strangeness was soon driven off, only Ronny kept pretty quiet. While the teasing and leg-pulling was going on, he let his big blue eyes quietly wander from face to face. He was a deep one.

"You talk too darned much, Ronny," cracked Art. "Once you get him going though, he won't stop, but I have never got him going yet."

"Shucks," said Ronny. (We later learned that was the strongest expression he ever allowed himself.) "You'll do the talking for two."

"Ronny comes straight from the farm," came back Art, "He only talks to horses and cows. I'll bet you, home was never like this."

"If it had been, we wouldn't be here," cryptically answered Ronny determined to keep his post a mystery.

"One never knows, does one?" said Art. He sure was trying hard to get Ronny going.

It was time to catch some shuteye. We had to double up in our pup tent that night. Stuart and I shared one.

The next morning, Stuart complained about bunking with me. "Holy mackinaw, you surely sawed them off last night, Hermann. Didn't get a wink of sleep."

"What, me? It was you who cut into the big ones! Didn't you hear me call 'Timber'?" I asked.

"Yea, that's right, I thought you had a nightmare, not used to sleeping under trees." Stuart could tease.

Somebody pulled the tent flaps apart: "Hey, you guys, stop arguing, get up and split some of the logs you cut last night. Time to get breakfast, there is work waiting. Tonight, you'd better move your tent to the point, so we'll all get some sleep. Maud thought there were bears around."

We consumed oatmeal mush, flap jacks and bacon, and lots of coffee and off we flew to safeguard our showing. "Let's have a look at it first, Hermann," said Pat. "We'll make a guess at the dip of the formation, that will determine the way we'd want our claims, to get sufficient depth. Here we are. Let me have your compass. Thought so: runs about true north and south. Same as the length of the Lake. Dip is about 40-45° east. You better make it three claim-lengths to the east, one to the west and six claim-lengths north and south."

"OK Pat. I'll look at the map and make two sketches, while you get the other two fellows. First post goes right here at the find! Want to witness the historic moment?"

"OK Then I'll be off. The sooner this gets done, the better."

Pat was gone and back again in short order. Stuart and I had a good look around the neighborhood, in the meantime. "You know about staking claims, Stuart?" I wondered.

"Sure do."

"Good… I figure the quickest way will be to work in two groups. Art and you, and Ronny and me. One can step it off and run the compass, the other one can blaze. By the way, Pat told me the variation is around 30° to the east of true north, so we'd better set our compasses for that."

"Got your sketch?"

"Yes, let's go. See you tonight!"

Ronny and I went to run our farthest line on the East side of the lake with the all-important No.1 posts. First south, then three claim-lengths east, which meant working our way through dense undergrowth up to the crest of the second range. Along the top the going was good, but in the draws, or ravines where there was often lots of overburden, it was hot and humid where the rain and snow got trapped, and clouds of mosquitoes travelled with us. That didn't bother Ronny much, mosquitoes didn't like his blood, the lucky guy. They made up for it by feasting on me; pine tar and citronella were only too soon washed away into shirt and underwear by streams of sweat. However, a fellow gets used to it and in a while the bites don't itch much anymore. Whenever we stopped to put in a post, I relit my briar. That helped somewhat, while I was writing my hieroglyphics.

After we had gone six claim-lengths to the north, we cut west toward the lake, and then made our way back over caribou trails which ran close to the shoreline all along the east side of the lake. We surely were thankful to the caribou for making travelling easier. Thousands upon thousands of sharp hoofs must have pounded along those trails, through many migrations, cutting cleanly through fallen trees, some of which were almost two feet in diameter when they fell into the caribou highways. It seemed somewhat mysterious to us. Then, later, after we had watched some herds pass by us for hours and days, nose to tail, we understood.

Pat flew our weary and begrimed bunch back to camp that night, and this procedure was kept up for a few days, until Pat managed to get another tent for us, and we could build our camp on a small rocky peninsula near a sandy bay, where Pat could run the plane right onto the beach.

While the four of us began clearing the first discovered vein of deadfalls, half-burnt trees and overburden, Pat and Reg were busy carrying more supplies to the base camp and spent the remaining time with us.

One morning Pat came roaring over our camp, while we were still in our bedrolls. Hardly had we pulled up our pants when the floats of the plane ground on to the beach.

"It's about time you guys got up. What do you think this is, a picnic?" hollered Pat.

"What's all the noise about so early in the morning?" I asked him.

"The ship called *Pelley Lake* is coming in today with the scows. I need two of you fellows to help move the stuff."

There followed two hectic days portaging and flying supplies to our base camp at Neely Lake. Our diet had been somewhat monotonous. Now we were going to live in style. Not only quantity but variety was assured. All day we lugged cases of canned goods, dried fruits, flour, coffee, tea, dried milk and eggs, bacon, and sausages in cans, enough to stock a fair-sized grocery store. Then there were pots and pans, dishes, and utensils. Shovels, picks, sledges, carpenter tools, nails and spikes, canvas, two more wall tents and a new bigger stove. Not to forget aviation gas and oil. We were so busy handling our stuff, that I had paid only scant attention to what else was going on at the portage. All of a sudden, a familiar voice hailed me. Wonderingly I turned round.

"Hello, there!" It came again.

"Hello! Oh, it's you!" exclaimed our young friend from the Muskeg Ltd., Ronald.

"Gosh, am I glad to see you!"

"So am I. How's it going? Getting used to the North?"

"Not so hot. Been sitting on this portage ever since we arrived."

"No prospecting? What did you come here for?"

"Lord only knows. All I've been doing so far is carrying water, cutting firewood, and getting the meals for the two old buzzards."

"So, you are not a bit happy, or are you?"

"I am damned lonesome, and sick and tired of this business."

"Well, come on over to our camp tonight. You'll like it. We got a great outfit, and everybody is happy."

So, we had a visitor, the first of the many who later were made welcome at our camp. Carrying a big bundle of Montreal Stars newspapers, Ronald joined our little party that night, his eyes fairly popping when he glimpsed our layout.

"Boy, this is swell! I wish I were with you. Want some Montreal Stars to read?"

"How about it, Pat? Couldn't you hire him?" I asked.

"Sure, you are hired. Stay here tonight. Get your stuff tomorrow!"

"Thanks a lot, but I can't do that. You see, my old man doesn't think I'll stick it out, and, by golly, I am going to show him. If it's all right with you folks though, I'd like to come over whenever I can."

"Sure thing, one more scrapper is always welcome in this gang!" Pat said.

First thing next morning we put up a bigger cook-tent, built a table from some packing cases, and organized the supplies for our camp at Neely Lake.

"Got everything, Hermann?" Pat asked.

"Looks like it. I'd like to have the little stove though, and then how about the dynamite?" I asked.

"We'll get that at B&M. Let's hop over now."

"Hey, Reg! Plane ready?" I was ready to go.

"Yep," Reg nodded.

"Put up the big stove in the meantime. Hermann wants the small one."

"OK."

When we landed at the B&M camp, Pat said, "Keep your eyes open. We are going up to the drill set-up and assay office. Look at the rock, pick up some samples if you have a chance!"

At the assay office, Pat met an old friend. "Well, I'll be a son of a gun, Pat. Never thought I'd see you again. What the hell are you doing here?"

"Meet an old friend of mine from BC! Hermann, this is Bud Jones, one of my old drinking chums. Went up the Skeena River in British Columbia together in '26. Can we get some powder from you guys?" Bill Windrum introduced us both to Bud.

"I'll take you over to Bill. He is in charge."

With Bud leading the way, we walked over to the powder shed. I kept my eyes peeled to study the rock. Finally, we came to a clearing where thousands of sticks of dynamite were spread out in the sun.

"Bud, what in hell is this?" exclaimed Pat.

"Scow ran aground and sprang a leak; the stuff got all soaked. They are trying to dry it out now."

"Hi, Bill, how are you?" asked Pat as Bill joined the group.

"Pat, of all the…. What brings you here?"

"Can we get a couple of cases of powder from you? But none of this stuff," he said, pointing at the spread-out sticks.

Pat explained, "I've got a bunch of greenhorns with me, and I don't want any accidents."

With a case of dynamite securely tied to my pack board, I picked my way down towards the dock along the edge of a barren ridge. Here and there I noticed irregular quartz veins cutting through the reddish porphyry. I should have paid better attention to the trail.

Slipping on some damp moss, my feet went out from under me. Down the hill I went, feet first, case of powder bobbing merrily over the rock, and me on top of it, trying to dig in my heels to stop the mad descent. Down, down, short seconds stretched into little eternities.

Finally, I came to a stop, crawled out of the shoulder straps of my pack board, walked shakily off a little distance, and had myself a smoke. A tear in the bottom of my breeks and a few skinned knuckles were all the damage. I was damned careful after that.

We flew the powder to the Neely Lake site for use later, then returned to base camp which was just being set up. Looking at our new cook-tent I saw Reg just tying the stove pipe to the end of the ridgepole. Then I looked inside, "Holy cats! Here's another guy trying to kill himself!"

Reg had made the job easy for himself. Instead of building a base for the stove, he had rolled two empty gas drums into the tent, an open bunghole right underneath the fire door. Off and on a spark was flying out of the draft slot.

"What's going on here?" said Pat.

"Have a good look," I said.

"Well of all the stupid so and sos! Hey, Reg, come here!"

"Careful with that stove you two. I've got a chocolate cake in there!" answered Reg.

"Chocolate cake be damned. Get those drums out of here!" yelled Pat.

"What for? They're empty."

"Get them out, I tell you, roll them down to the beach, and I'll show you."

Off to the beach we all went and Pat threw a lit match into the bung hole. WHAM!... The drum jumped a foot into the air, toppled over and rolled into the water.

"No danger, eh. God bless my teeth, but how in hell did you ever manage to survive this long?"

"Well, have it your way! I only hope my chocolate cake didn't collapse. This is Sunday, you know!"

"What?" Pat asked in a louder voice.

"Sure, it's Sunday!"

We had completely lost track of the calendar. From sunrise to sunset, we were on the go, so much to be done, every night we had a full program for the next day. No whistles blew, no clocks were watched.

Reg's chocolate cake was all right, and we enjoyed it, but Sunday or not, we decided to haul some of the new supplies to Neely. I thought of staying there, but Pat had another idea. He had noticed a very attractive lake with a long sandy beach somewhere north and east of Beaverlodge.

"We will go over, have a swim and do a little prospecting. Maud would like that. What do you think?"

"Anything goes with me. I don't think we should stop prospecting just because of the first find. After all, how good is it?"

"Certainly not. Neely is safe from other prospectors now. We got to keep going." With the claim stakes in place and still about two miles from our base camp, Pat began circling to land near the north-west shore of Beaverlodge Lake. I was wondering what was up when his voice came through the speaking tube.

"There is a fellow with a skiff down below. Maybe that Swede trapper. We'll call on him."

As Pat had seen, I spotted the skiff with a kicker pulled up on the shore. We landed, but no human being was in sight, although a fire was still burning, and a full coffee pot was standing nearby.

"He doesn't want any company, Pat. He ran away from us!"

"Must be bushed. We'll sit down just the same. Like to know what kind of neighbors we've got."

We sat and smoked, facing the lake, acting as if we didn't expect to see anybody. Ten, fifteen, twenty minutes passed. Nothing happened. Half an hour. Finally, we heard soft footsteps behind us. Pat turned around.

"Hello, there! Your coffee is getting cold."

The poor fellow looked at us as if he were going to take to the bush again. He sure was a shy specimen. Pat had to turn on the old charm to get a few words out of him, even after he had offered some of his coffee. That's the way we met silent, mysterious Per Larson, the trapper.

"Come on over and see us some time, Per?"

"I do dat," he promised, but we hardly believed him. He didn't seem to care for human company.

On the step of the floats, we raced to our camp, Maud and Reg looking out for us.

"Anything wrong with the plane?"

"Not a thing. Just called on our neighbor, a silent Swede!"

The sun was still high, when we reached Pat's swimming hole, a gem of a lake. Wide sandy beach, facing south, scattered jack pines in the background, a gentle breeze rippling crystal-clear water.

"What a place for kids!" Pat remarked, and I thought I detected a dreamy look on his usually slightly sardonic countenance.

"It would be nice to have our children grow up here in God's own country!" I replied.

"Our children? You haven't even got a woman yet! Or are you holding out on us?"

"No. But I can dream, can't I?"

CHAPTER VII
WHILE THE CAT IS AWAY

THE WEATHER CONTINUED perfect, for a week or so, and work was progressing well at Neely. Dynamite cleared away the upper layers of rotten rock in our trenches, and in the solid rock more and more mineral showed up. But no free gold. However, Pat assured us that often the highest assays came from rock where the gold was so finely distributed as to be invisible to the naked eye.

"Let's take some grab samples from the bottom of all the trenches and mark them well. I am going to take them to Edmonton to have them assayed. While I am gone, I want you, Hermann, and Stuart to stay at the base camp."

A lot of speculating was done that night at Beaverlodge camp.

"I'll bet you we'll have the diamond drills going before long if this stuff kicks. I've seen ore before, this stuff looks good, I tell you." Pat was very confident. "I'll buy you each a bottle of Scotch if the assays are no good. If they are good, you each buy me one. How's that?"

"OK with us!"

"What kind do you want, Hermann?"

"Johnnie Walker, Black Label."

"Make mine Catto's."

"Christ'a'mighty, but you guys are fussy. It will give me great pleasure though to make you pay for some Haig & Haig, or Buchanan's."

This thirst-raising conversation was unexpectedly interrupted by the chug, chug of an outboard, motor coming closer and closer.

"By golly, it's the Swede! Who would have thought that?"

"It's that irresistible charm of yours, Pat. He's making straight for here."

Sure enough, nearer, and nearer he came, then cut his motor, tipped it up, and let his skiff ride onto the beach.

"Holy cow! Look at that stuff, Hermann," remarked Pat out of the corner of his mouth, "that guy is not visiting, he is moving in!"

"Hi, there Per! You moving?"

Per said "Hello" and not much more and stayed.

Early next morning, Wednesday it was, Pat and Reg and our samples went off to Edmonton. "We'll be back Sunday." Pat said, "Don't start worrying before Monday night. Keep your noses dry!"

"Wonderful piece of advice that is," I answered him, "What if you are not back by Monday night?"

"I'll get word to you if we are held up."

"All right, Monday night we start worrying, because you fellows couldn't get out of the beer parlor in time!"

"Contact."

"Contact."

"There goes your birdman, Maud," I said.

To fill the time, I had a suggestion. "Now to make the days shorter, we've got to do something. Stuart and I were talking about cutting a trail along the shore to the rocky point over to the east. It's a nice breezy spot, not so many mosquitoes, and we might do some fishing there." Per had gone off in his skiff on some mysterious errand, and returned to camp, however, about the same time we did. This time he was towing a canoe behind his skiff, both vessels heavily laden with a wondrous assortment of things. Apparently Per had decided to throw his lot in with ours. Nothing much had been said, so far.

However, Per had shaved and changed into new clothes while away, and we figured this was telling us enough. His canoe was a most welcome addition to our equipment. It gave us a chance to do some exploring along the lakeshore, and to visit some of the many small islands dotting the lake.

Sometimes Per joined us, but more often just Maud, Stuart and I went on these trips, which taught us many a lesson. The north side of the islands was covered with glacial drift, a wonderful and interesting collection of rocks. Here and there along the shore, we found weather bleached teepee poles, neatly arranged like the spokes of a wheel, just as the Indigenous men had left them when breaking camp, silent witnesses of their wisdom. They always built their camp, at least in the summer, on an island or point exposed to the prevailing wind, which kept the flies and mosquitoes away.

Thus, the days went by quickly, Sunday came. Off and on the sound of planes was heard from the direction of Beaverlodge Bay, but no Pat appeared. Monday came and went, Tuesday, Wednesday, no plane. The strain was beginning to tell on Maud, and Stuart and I began worrying too. We decided to go to B&M camp, where they had a radio sending and receiving station. Per was busy fixing his outboard motor, so we took the canoe across the portage, and paddled towards the big camp. Lake Athabasca's waters were smooth and peaceful. Sparks, the radio operator, didn't take us very seriously at first, "Why worry? They are probably having a hell of a swell time in Edmonton. I wish I were there."

"Look, Sparks, do us a favor and find out. Pat was supposed to be back here by Monday night at the latest. This is Thursday!" Stuart was adamant.

"OK boys. But you'll have to wait until 5 p.m. I'll try and get Edmonton then. It's no use now, they are not listening on our frequency at this time of the day."

Stuart and I looked at each other. "5 p.m.? It's 10 a.m. now. We'd better return to camp and come back here tonight." It was another futile trip that night.

"No sign of Pat," Sparks told us. "They tried the MacDonald Hotel. He isn't there."

"What about Cooking Lake? Surely the plane must be there?"

"I got this via Trail. Couldn't get hold of Edmonton directly."

There wasn't a very cheerful bunch at our camp that night. As long as daylight held out, we expected to hear Pat's motor any minute, but no such luck. Then we tried to find favourable explanations for his failure to show up. Next morning Per proposed to use his reassembled kicker and the canoe to go to B&M. A canoe with a kicker attached is a sight to behold.

"No news," again came the disheartening answer. Per who had gone to see the local trader, picked us up, as usual not saying much. Stuart and I were swearing profusely to vent our feelings.

A few minutes later our minds were occupied with a different matter. Hardly had we left Beaverlodge Bay when a sudden wind-squall hit us square on. Waves turned into whitecaps. Stuart, sitting up in the bow was drenched in no time. We began to ship water. In spite of Per's efforts to keep the boat at a right angle to the waves, I was soon up to my middle in ice cold water, bailing for all I was worth.

Per was a master skipper. Without him it might have been a cold end for us. Holding the groaning and creaking fragile craft straight into the wind he was heading for the leeside of a big island, whose outline off and on we glimpsed through the flying spray.

Bigger and bigger grew the waves. When we got onto the crests, the bottom of the canoe came up, as if it were going to break in two. Down we went into the troughs, kicker racing, Per working the gas lever and the steering gear desperately, until the bow came up, and the propeller bit in again. After what seemed eternity, Per got us into calmer water close to the island.

As suddenly as the wind had blown up, its fury now abated, and over placid waters we rode toward the portage. Cold and shivering in our wet clothes we stepped once more onto terra firma, legs somewhat wobbly. We were safe, but still we had no news for Maud.

"Let's leave the canoe on this side of the portage, we have to go back to the radio station this afternoon," I said to my chums. Nobody objected. It was tough having to arrive at camp without news. Already the peril we had just escaped was pushed into the background.

After changing into dry clothes and having a hot meal, and we were ready for more action. Per stayed in camp, while Stuart and I once more crossed the portage this time with just the canoe.

"We will stay there until we know," we said before setting out. "If the Lake is too rough, we'll hike it. So don't worry about us! We'll bring good news."

The big lake was calm, as if it had never seen a storm. A soft balmy breeze barely ruffled the glittering expanse. Sparks, the radio operator, must have seen us coming, he stood by the door of his shack. "Pat's still in Edmonton, plane at Cooking Lake. Hang around a while. I have another schedule. Won't be long now."

Finally, finally, it came: "Left Cooking Lake this morning. Beaverlodge tonight. All is well." We never paddled so hard before, but progress seemed slow, so slow. Over the portage we raced, canoe on high, and when we rounded the point and the camp came in sight, we sang as loud as our wind would permit. "Happy Days are Here Again!" and waved our paddles. "Tonight, they will be here! Tonight!!"

"Thank you, boys, thank you!" was all Maud could say, and there were tears in her eyes. Poor girl. She had had a hell of a time. Now there was joy and relief. Nothing unforeseen had befallen crew and plane. Of our samples, the reason for the trip, nobody had thought, and nobody was even thinking now. Soon we would know the explanation for the delay, but even that wasn't much in our minds, now so happily filled with expectancy. Several times we heard motors. "Listen, here they come! No, damn it, that's a kicker."

Then again. "Here comes a plane." Plane all right, but not Pat.

So went the afternoon. Evening fell, and our camp was already in the shadows of the range that rose to the west. Semi-darkness was stealing over the lake; only the north and east shores were still bathed in sunlight. We had almost decided to give up waiting, when seemingly out of nowhere, we hadn't heard the sound of an approaching motor, but Pat's plane came roaring over the camp at tree-top level, but not to land immediately.

Up, up it went in a steep climb, till it seemed like just a glittering moth, caught by the last rays of the setting sun, way up high over the northeast of the lake. Diving and climbing, looping, and rolling, Pat surely took her through her paces, and we, the waiting, stood by the shore close together.

"Son of a gun" squeezed out Stuart between clenched teeth.

"Pat, oh, Pat," was all Maud could breathe, gripping my arm tightly.

"Relax my dear, your Birdman is coming home to roost." And then at last, the falling star turns into a plane again, the rigging screams in a sideslip, and the wandering-bird hits the home waters, making a final little curtsey, as floats are touching the beach.

"How goes the battle?" asked Pat, hoisting himself out of the cockpit.

"Battle is right, brother, but it's just starting!" I felt like throttling him.

"What goes on?" Pat looked puzzled.

"You are only about a week overdue. How about some explanations?"

"Hey, Reg, didn't you send the wire?" asked Pat, still puzzled.

"Sure did."

Well, we never really found out, but the bedrolls told some of the story. Said Pat, "Your Scotch is in my bedroll. Samples were no good, all traces, no real values."

Stuart and I made for the plane's cabin to get our prize; a stiff drink seemed a damn good idea.

Just then the door opened and out crawled a fellow with a greenish-gray face, on unsteady feet balancing along the float toward dry land. Shaking himself, then stroking his fevered brow with a limp paw, the apparition just murmured.

"Lordy, Lordy. Where … am I??"

Pat, busy with explanations for his spouse, swung around.

"Almost forgot! Meet Don. Met him in a beer parlor. Did I shake you up too much, Don??"

"Be all right in a minute. Are we home or did we just land for another beer??"

Oh, yes, the scotch, out came the bedroll, but what came out of it? Cigarette butts, empty packages, bottle tops, ash trays and burnt matches, empty bottles, a towel and… a Gideon Bible. "What a party that must have been," remarked Stuart, finally pulling out the eiderdown.

"Easy now. We must be close." Sure enough. There they were, gurgling invitingly: Catto's and Johnnie Walker's. It was said afterwards that Stuart and I sang beautifully but we greeted the new morning, how quickly night had flown, in remarkably good shape, although off and on my brains seemed to bump against the walls of my skull.

As a most welcome surprise Pat had brought a battery radio, and some wire for antenna and ground. Everything was in astonishingly good condition despite Pat's antics in the air.

So, after we had our breakfast and plenty of coffee, we set to work to install the antenna. Our aim was to get it strung clear of the surrounding bush. Carefully we picked the two tallest trees near our tents on approximately a north and south line. Then we strung out the wire.

"Where are the insulators, Reg?" asked Pat, "or didn't you get any?"

"Sure did." Frantic search. No results.

"What's the matter with beer bottles? They should do the trick." was Stuart's suggestion.

"How are you going to get the wire up, Hermann?"

"Climb the trees and chop off the branches as I go up. I'll tie one end of the wire to my belt. Here we go."

That procedure worked all right. Leaving about a foot length of the stronger branches sticking out from the trunk, I assured an easy descent. After all, it was no great undertaking, and the only reason I talk in detail about this is the fact that we were really trying to assure a good functioning of our wave catcher. Quite proud of our handiwork, we then pulled our lead over to Pat's tent where the receiver was put up.

"Bet you, we'll have wonderful reception. No streetcars, no static."

This was going to be good.

"OK Pat, give her the juice."

A little hum, crackle and hissing, that was all.

"Tubes all OK?"

"Yea."

"Ground connected?"

"Yes."

"Batteries?"

"OK"

"What in hell is the matter with the blasted thing?"

"The plane ride? Maybe the speaker is still dizzy?" I consoled Pat. We interchanged the ground with the antenna. No results. A downhearted bunch viewed the Marconi radio receiver.

"Only one more thing we could try: change the direction of the antenna," I suggested.

"You got the damn thing up there; you know how to get it down again."

Up I went and cut the wire around the tree.

"Watch out down below, here she comes."

The beer bottle insulator hits the ground, and out of the tent floating up to my lofty perch are coming the haunting strains of the "Rhapsody in Blue."

"Hey, up there. She sings!"

"Yea, sounds fine up here too."

I don't know any explanation, but our radio sure worked fine with the antenna resting on the undergrowth, here and there touching the ground. For a short while we enjoyed the long-missed pleasure of listening to music, but it was high time to return to our chums at Neely Lake. We, too, were overdue.

After lunch Pat flew Stuart and me to our other camp, where we found Ronny and Art in good shape and good humor. They had done a lot of work on the trenches. Ronny was developing into a real powderman (he just liked the stuff). It was good to see him happy with his achievement.

"Gives me the creeps, though," said Art, "to see him bite the caps." Fortunately, I had asked Pat to get us some crimpers and a handbook on the use of dynamite from Edmonton, and these I could now hand to Ronny.

"You boys got enough grub here for a while? Anything else you need?" asked Pat. "I am going to take Don for a trip to Fort Chipewyan. Be gone about a week."

"We've got enough for about two weeks. Only don't forget to bring us some more tobacco on your next trip, and how about trying to get us a canoe. It would save us a lot of walking, especially when we start prospecting again."

"All right boys, keep your noses dry. I am off to see old Dick. If all goes well, I'll be over tonight to bring you your tobacco. And for Christ's sake get busy and get some better samples this time, or I'll fire the whole goddammed bunch of you!"

"Yes, sir," said Ronny quietly, and in an aside just as quietly, "Strong language he uses, doesn't he?"

"What's that?" came from Pat.

"Yes, sir!" Ronny responded.

Pat didn't show up that night. A storm broke the spell of fine weather, and rain was hammering on the taut canvas of our tents next morning. Dense fog was shrouding the landscape, visibility zero. Soon the floors of our tents were a quagmire. Everything felt clammy. We hadn't spent much time at our camps so far, we were too eager to get the trenches across our veins blasted out. If we wanted at least our bedrolls dry, we had to act quickly. Just above the old camp we found a solid rock plateau where we hurriedly put-up log walls about three feet high and erected the tents on top of these. Fastened to the log walls we built frames from poles, stretched gunny sacks over them and presto, our beds were ready.

The rain kept on pouring, down, off and on there appeared a little break in the skies, but always the fog came rolling in again, and then there was more rain. All day long we kept our little tin stove going, talked a little, read a little, and smoked too much. It kept on raining, day after day.

"Tomorrow," we said, "tomorrow the sun will shine again." More rain. Then our tobacco supply began to run low. We pooled our resources and divided them equally, a pitiful heap for each one of us. Before another day was half gone, there wasn't any left.

"A cold man's comfort, a lonely man's companion." We learned the full meaning of those words.

"Why in hell did I ever leave home, to come to this God-forsaken country?" lamented Art, "not even a smoke, and bully beef every day!"

Stuart, who had spent more time than any of us in the North quietly answered, "We all say that at times, when the going is a little tough, maybe, but mark my words. We all come back."

"Well, you wouldn't catch little Art doing that. Next time I get ideas like that, I'll think of steak smothered in mushrooms at Johnson's. That will make little Art stay home."

"I don't believe that," Ronny used one of his favorite expressions, there was something very final about that.

"They say you can smoke tea leaves," Stuart said.

Art made a heroic start, filled his pipe with tealeaves, dried over the tin stove. We followed suit. But not for long. Choking and wheezing we put our briars down. It would be easier to give up smoking altogether we decided. But, after a few hours, we fell victim to another one of Stuart's suggestions. "I've heard of guys smoking moss."

He shouldn't have said it. We tried it. "Maybe we will get used to it??" We tried for three days. We didn't believe it. It must have been nine or ten days until finally one evening it began to clear in the west and we saw at least the sun setting, and the following morning we awoke early to an unusual brightness.

The rain had ceased. As the sun marched across the sky, steam rose from the drenched country, and with it clouds ... of mosquitoes. In the sultry atmosphere the mosquito dope didn't stay very long on our faces when we started on our trenching operations again. Now we didn't even have anything to smoke to keep the buzzing pests away. We built smudges to make things bearable.

"Goddam buzzing drives you nuts. Give me a nice cool beer parlor on Jasper Avenue, or did I mention that before?"

"You did, Art, but wait till the black flies come out. This is nothing yet," answered Stuart.

Again Ronny came in with his final, "I don't believe it." What relief when toward evening Pat came breezing in. The whole bunch of us ran to the landing.

"Got a smoke, Pat?" A package came flying before he got out of the cockpit. "Couldn't make it sooner, couldn't take this crate off the ground. I got back from Chip just in time, before the weather closed in."

"Any mail?"

"Got it this morning. Help yourself, it's in the cabin. By the way, the big boss is coming in to have a look at our show. A lot more prospecting parties have arrived, most of them are just sitting around, waiting for something to break. Only our party has a plane though. One old prospector told me he is heading this way by canoe. Pete Lauder is his name. He is supposed to be good. What do you think, Hermann, have we staked enough claims?"

"Well, that's hard to say, but remember I told you about some mineralized veins just outside our eastern boundary. If this fellow you are talking about comes from the east, we'd better stake another row, that would take us into the valley behind the last range."

"You better start on that tomorrow. I am going to Chip to pick up Don, told him to look around for a canoe."

"That'll be a help. Here is a list of other stuff we need, and if you and Reg have any more smokes on you, better leave them here."

"Holy cats," said Pat, glancing at the list, "what do you guys think this is? A summer resort?"

"The only thing that could have reminded us of such places, was the rain of the last ten days."

"All right, I'll see what I can do for you. Here are some tailor-mades to tide you over. I'll be on my horse. Get those claims staked and when you find the time, why the devil don't you guys shave?"

"Yes, sir," said Ronny, whose reddish-blond whiskers were beginning to curl.

"Bring us some blondes, and we'll shave every day!" Art called out, stroking the black, wiry growth on his chin.

"You'll shave, or I'll fire you. Plain insubordination, that's what it is."

"Keep your nose dry," I called after Pat, who crawled into his cockpit, grinning.

Along the caribou trails we pushed next morning to stake more claims. From the southeast corner post one claim length to the east, then north again. Two of us went that way, the other two followed the old boundary line and cut the lines running to the east, to the posts we put in. 1500 feet doesn't sound like a great distance, it's surprising though, how easily you can miss out on hitting the posts on the nose.

Following our compass, we soon descended into the sandy valley, and suddenly we came upon a spot where somebody had camped. Some empty cans, not rusted yet, were lying around, some tent pegs still sticking in the ground, and some blackened stones surrounding a heap of ashes, were witnesses of a human being's presence, not long ago.

"Must be that old prospector Pat was talking about."

Before nightfall, our claims were staked. Hardly had we reached our camp when to our surprise, Pat came in again, just to bring us some tobacco.

"Seen any signs of the old bush rat?" was almost his first question. "Claims staked?"

"Claims are staked, that is, we got all our No.1 posts in. Saw signs of a camp near the creek east of our claims, must be your friend," I answered.

"Friend, be dammed! Let's meet the guy first. What are your plans for tomorrow?"

"We'll run the last lines, then come back to camp around the north end of the Lake, prospecting as we go. It's pretty rugged up there, lots of breaks in the formation."

"Good idea. Watch for the breaks and shear zones. I'll be on my way. Good luck." Pat was gone.

Late the next afternoon, after all our lines were blazed, we suddenly ran into another freshly blazed line. We looked at it the way Robinson Crusoe must have looked at the footprints of Friday. After plowing through virgin wilderness all day, it was almost like a shock to see a sign of human activity even though we knew someone had been around the area based the camp by the creek. We followed the line till it petered out.

"That's not a claim line. No posts. Somebody just marked himself a trail." Stuart suggested.

I looked at my dollar watch. "Too late to investigate further today. Let's get back to camp. When we get the canoe, we will find out."

The canoe arrived a few days later tied to the undercarriage of the plane. Pat had carried it all the way from Fort Chipewyan. It wasn't much of a craft, a miniature edition of a canoe, a rat boat, about ten feet long with a very narrow beam. To Pat's chagrin we voiced our disappointment.

"No good you guys say? I'll show you."

He showed us all right. When he climbed into it the water came up to within three inches from the gunwales, and it was almost into the drink with Pat.

"You'll have to get used to it, that's all there is to it." And that was that!

"OK! We'll improve on it a bit."

Not only a canoe had arrived, but a new member for our crew: Scotty, formerly with the Hudson Bay Company, a tall, skinny, sandy haired true Scot, who got tired working peacefully for little for the Company of Adventurers.

Ronny and Art tried out our canoe.

"No damn good," was their verdict. "That thing was built for one hungry guy looking after his muskrat traps."

"Let's put a keel on it, and some cross-braces."

With axe and hunting knife as our only tools we worried a keel out of a fallen tree, likewise, cut braces for the fragile craft, and made some double paddles out of sticks and boards from a bully beef case. After that, as long as the lake was calm, we even rode it with three men in it. Two men paddling, one man bailing! However, it saved us a lot of walking.

One day three of us went across to the east side to follow up the outlet of Neely Lake to do some prospecting. We were deep in the bush when we heard Pat coming in and leaving again after a short stay.

"The big boss was here!" The other boys reported on our return.

"Well, what did he say?"

"He likes the show. Talked to Pat about diamond-drilling and a winter camp."

This was big news. The day after, Pat confirmed it. We were all very happy about it, and immediately started planning for the winter camp.

"How'd you like to spend a few days at Beaverlodge with us, Hermann, we can plan the details there. I may have to go to Edmonton again about supplies," Pat explained.

"That's fine with me. But for the love of Pete don't talk about going to Edmonton, my nerves couldn't stand it. Say, whatever happened to Dick? I have forgotten the old grouch over all this excitement."

"You know the old so and so threatened to shoot me last time I went to see him! He is bushed completely. Took him to Beaverlodge. I think the big boss will send him back to his beloved Ontario, Northern Ontario, that is."

On our way, over to Beaverlodge camp, Pat knocked on the little window between cabin and cockpit, pointing to the speaking tube. I grabbed it.

"Look who's coming, to your right, down below," came Pat's voice. "Hold on tight. I am going to scare the pants off him."

I just had a glance of a small biplane below, and then, down we went in a power dive straight for the little fellow. I thought I saw the whites of the eyes of the pilot in the open cockpit. Just before Pat pulled up in a steep climb, banked into a sharp turn, and down we went again, dove under the other guy, up again. Then I lost track of our movements. One moment I saw only the skies above, the next the lake, or the ground below racing toward us at the craziest angle. Holding on for dear life, I closed my eyes, while Pat enjoyed his aerobatics. My stomach felt as if it had been torn from its moorings, and didn't know which way to go, through my clenched teeth or the seat of my pants.

I began to feel the way Don had looked when he first came to our camp. Finally, we slid down towards camp, and high time it was too.

"What's the matter, Hermann? You sick?" asked Pat.

"Not at all, I am fine now. Just pushing my innards into their normal location again. What happened to the other fellow? Who was he? Poor guy."

"Fellow by the name Steve something. He just kept on going. What else could he do?" chuckled Pat. "By the way. I've got a surprise for you here."

"If it is a drink that would be most welcome," I responded.

Suddenly, a familiar voice greeted me. "Oh, hello, Hermann. How are you?"

"Fine, just fine, thanks Maud. You are looking well. Great life, isn't it?"

Pat was looking around for something, then turned to Reg, "Hey, Reg, what did you do with that surprise package I picked up for Hermann?"

"Must be asleep somewhere. That's all he's been doing since you went."

"Holy mackerel! Go and find him."

Reg appeared with a seedy looking specimen, who quaked in his boots when Pat roared at him, "What in hell do you think this is, a vacation spot? Sleeping in broad daylight, I'll be damned! Hermann, here is your man Friday, a fine-looking representative of the Teutonic race, isn't he?"

"Do I get that drink now, or don't I?" was my answer.

"Need it?" Pat asked.

"More than ever now," woozily the words came from me.

"As I always say: No sense of humour!" Pat thought he was being funny.

"OK I know the Irishman's lament by now. We really have to go into that, some long winter evening."

"Well, here's looking at you." Pat grinned.

"Prosit," I offered a toast to Pat's health.

"Whatssat?" asked Pat and the irony had to be explained.

The fun over, we went to work, checking supplies on hand, making long lists of items needed, and the long, long list of groceries, enough to feed 25 to 30 people, if possible, until open water next spring. The camp had to be built in a hurry. More men were needed immediately.

"Where will we put the camp, Hermann? You know the ground best by now."

"For the winter we want shelter and firewood. We'll find that only at the east side of the lake. You must have noticed a T-shaped peninsula opposite our present camp. Looks to me like the best spot, somewhere at the base of it. Fairly level ground and thick bush."

"OK. We'll look at it when I take you back tomorrow, before going to Chip," Pat suggested. A sense of excitement was building. It would be a long winter with lots of work ahead of us but we had a plan.

CHAPTER VIII

WINTER CAMP AND FREEZE UP

WE DECIDED ON that spot at the t-shaped peninsula. Every morning our little crew crossed the lake to clear the ground for our buildings. Sawing, chopping, and dragging, we labored from early to late, 14 to 16 hours a day, to bring order into veritable jungle. All usable timber was limbed and piled up, the undergrowth cleared away, by and by a large enough area took on the looks of a park.

Part of the timber we had cut came in handy when we started building a dock for the planes that would bring our supplies from Beaverlodge. Pat had been busy making all the arrangements for the supplies and their transportation. Off and on he would drop in, bringing supplies, mail, and new members for our growing crew.

We welcomed Dan, our new cook, on the scene. He was very welcome. Quiet, somewhat shy, and very likable, he was not very big (sometimes called Little Dan) but a tireless worker, who fitted well into our little gang, and that means everything in the bush. So far there had been nothing to disturb the complete cooperation of all. If one fellow did the cooking, the others would see that there was enough water and firewood and would wash the dishes afterward. Whoever thought of doing something for the camp, just went ahead with it, and always found helpers. We worked hard, driven by no one.

"Friday", however, never fitted in, and fully justified my worst premonitions when I first laid eyes on his seedy appearance. We couldn't move that guy with a stick of dynamite. He just couldn't get a move on, or wouldn't, and by and by we all became convinced that he wasn't quite right in his upper story.

Shortly after he arrived, he had dug out of his packsack, which was filled with a rare assortment of odds and ends, a good mouthorgan, his prize possession. We expected great things. He blew a few bars, vaguely reminding us of some song. Art, who usually was up to some tricks, looked at Friday with a rapt expression, "That's damn good, Friday! What's the name of the song?"

"It's the circus song!" exclaimed Friday.

To the end of my days, I shall not forget it. Night after night, Art with a deadpan face would say, "Give us a tune, Friday."

It never failed, it never changed. Two bars of the circus song, over and over again. By and by it got on our not very delicate nerves, but it was Friday's sleeping bag that finally finished his career with us. Ronny and Art had repeatedly commented, in very pronounced terms, on its strong aroma. When we started out in the mornings for our days' activities, they would stay behind and drag the ugly bundle into the open air to let the sun and wind go to work on it.

Friday didn't like that at all and stopped talking to us. Then one night Pat came in late. We had had our supper. Friday had gone off by himself. We sat in our tent discussing things with Pat, when suddenly from the other tent burst forth the, to us, so familiar dirge-like bars of the circus song.

Pat stiffened, "What in hell is that?"

"That, oh, that's Friday. He is playing just for you."

"Cripes, for a moment I thought I heard the wail of the banshee. I am going to stop that," we were happy to hear from Pat.

Pat strode over to the other tent, lifted the flap, stuck his head in, but jumped back immediately.

"Holy mackerel, that's awful. Pheeww."

"Do we have to say anything more?" I asked him.

"No, thanks. I see what you mean."

Exit: One man Friday.

When the first bunkhouse was finished, we abandoned our old camp. Several new men arrived, and construction of the cookhouse was started. We made it big enough to hold all food, supplies, as well as some bunks for the cook and the bull-cook who looked after camp chores like water and firewood. After the arrival of our winter supplies at Beaverlodge Bay, there was the roar of planes over our camp from morning to night. We were putting the finishing touches to our enlarged dock one afternoon when the first Canadian Airways Junkers arrived, her capacious belly filled with crates and boxes, and building lumber lashed to her under-carriage.

Hardly had we pulled the big machine alongside the dock, when the cockpit window opened and a stentorian voice bellowed, "Get busy you guys, and get this damned crate unloaded. We haven't got all day!"

"Who are you to give orders around here?" I hollered back.

Just as quickly came the answer from the pilot, "Who in hell are you?"

"Same to you. Who in hell are you?"

A big fellow climbed out of the pilot's seat, ran along the wing, and jumped right in front of me. We looked at each other, an engaging grin crept over his keen face, a hand shot out, "My name is May."

"I am known as Hermann around here."

Thus, I met the well-known Wilfrid Reid "Wop" May, hero of legendary exploits in Northern bush flying and a leading member of the knighthood of flying pioneers. With single motored planes, without radio, just with rudimentary instruments, they flew in the immense Northern wilderness. They hauled the supplies, carried the trappers and prospectors to any spot these fellows pointed out on the map. Any lake became their landing field, on pontoons in the summer, on skis in the winter, but in spite of primitive equipment, their safety record was superb. They were born flyers. Heroes all of them.

Days of unending activity, building, hauling, and storing supplies driven by unyielding urgency. Winter couldn't be far off, nights grew chilly, the Northern Lights became more brilliant, and there were mares' tails in the sky. When would the snow fly?

Maud and Pat moved into their new residence, a big tent with a board floor, and four-foot board walls. The cookhouse was finished, another bunkhouse was under construction, also a log cabin for the diamond drill foreman near the site of our original camp. Soon we would be safe, and it was high time too, the first light snow had fallen and vanished. Next time it might stay.

Talking about winter in the bush one-night Don and I decided that a small log cabin would be a better home than a somewhat reinforced tent. I took up the matter with Pat and he said, "So you think you can't take it, eh? A tent is good enough for me!"

"OK So, you are tough, we know that. All I want to know is: can we get a little assistance to get the logs down?" I expected a positive response.

"Where do you get enough logs?" Pat always needing the details.

"Halfway up the lake. We want to float them down, before the ice forms."

"OK. Help yourselves. Still think you are damn sissies," Pat said, ever the tough guy.

"Who laughs last. Ever hear that?" I couldn't resist.

Winter came in earnest as soon as we had our logs at camp site. Quickly the walls rose, and so did our confidence that we would live through the winter in comfort. Bob ably assisted us in the log work. He was an experienced guide, who knew the bush and could swing an axe. A big, keen, friendly fellow, that was Bob, a real guy, and a true friend.

Pat was still praising the comforts of his new home, but Maud often came over to see the progress of our cabin, and there was a wistful look in her eyes.

"How do you like it, Maud?"

"I think it's going to be nifty," she answered. "A good place to have a baby." She seemed to be mentally comparing her house to ours.

A few nights later, we had just started nailing the floor down I said to Don, "It's coming along all right."

"Yea, it's not bad at all, but I don't think we'll ever move into it," Don responded.

"I know what you mean, and I am glad you brought up the subject. We might as well settle it right now."

So, we trudged over to Pat's flimsy mansion. It was very comfortable, no doubt, as long as there was a roaring fire going, but the real cold hadn't come yet. Many an evening we had spent with Maud and Pat, talking, reading, listening to the radio, so that night too, Pat welcomed us, "Come in, make yourselves at home, take a load off your feet."

"We came over on business tonight, Pat. Don and I talked it over. We want Maud and you to take the log cabin."

"I wouldn't think of it. What's the matter with this place?"

Well, it ended by me talking like a real estate agent, extolling the virtues of our log cabin. Pat put up a good show of sales resistance. After all, hadn't he talked a lot about being tough?

"You mean to say, sir," he sputtered, "we can't take it? I spent many a winter in a tent!"

So, it had to be said, "We are not thinking about you alone. We know Maud would prefer the cabin. It will be more like a home, and it will be warmer. Furthermore, you people expect a blessed event, a delicate little person, and you know famous men are born in log cabins. So, this is just a friendly suggestion!"

"OK. Fellows, it's a deal. Thank you." The couple, expecting the first baby in camp, agreed to trade cabins.

So, it was settled. Said Don when we finally left, "Didn't think we would have to work so hard to talk ourselves out of our nice and warm little cabin. It's better, though, this way."

Shortly afterwards, Maud and Pat moved into the cozy little cabin. Brightly painted Scandinavian-styled cupboards and shelves, which I whittled out in my spare time, colorful curtains, and some caribou skins on the walls made an inviting home.

At night, we would gather there often and by the glow of the Coleman lamps, read, listen to the radio, and talk often into the wee hours of the morning.

The diamond drillers and their equipment arrived. The greatest rush was tapering off, and somewhat of a reaction set in. It got a little bit too quiet around the camp, and a sort of moody atmosphere crept in. Then one night, after the second canvas bunkhouse had just been finished, most of us were in the cookhouse having our supper, when someone cried outside: "Fire! Fire!"

The whole camp site was lit up by an eerie, flickering, brightness coming from the newly finished bunkhouse. In no time at all a bucket brigade was formed, but nothing could save the big tent. I don't know how he got there, but there on the ridgepole was Pat, last link in the chain of bucket-passers pouring water on the blazing canvas. It was all over in a few minutes, tent and fly gone, and a few of the bedrolls scorched and soaked, the framework blackened, but still usable. It was a loss, but the feverish activity of those few minutes acted just like a good tonic. Gone was the lassitude. Everybody was quite happy, believe it or not.

Each succeeding morning there was a wider rim of ice in the bays. The plane had been pulled onto the ground. Reg and some assistants were exchanging her pontoons for skis. Winter had set in for good. Freeze up time. No planes from the outside would come in for several weeks, until the whole country south to Edmonton became icebound. On our lake ice was soon stretching from shore to shore.

"A few more days and we can fly again," Pat was saying one night, while we were trying to listen to the messages sent out by the B&M station at Beaverlodge. Reception was often not very good, but that night it came through clearly, "Bill Windrum overdue on a trip from Prince Albert to Beaverlodge. Heavy fog over Lake Athabasca. Ceiling zero, visibility poor."

"Must have run out of gas trying to get across the lake," someone listening suggested.

"Let's have a look at the map. I've got to look for him," Pat said right away. "See this here? Archibald Lake, just south of Athabasca. Bet you anything that's where he is sitting."

"You'll have to wait till the ice is strong enough, Pat, and how will you know you can land wherever you may spot Bill's plane?" I asked.

"Well, you and Bob can go out to the middle of the lake tomorrow morning and measure the thickness of the ice there. Bet you it's three inches now, I need at least four inches for a safe take-off and landing. If it's not safe to land on Archibald Lake, they will wave me off. Where is Reg?"

"Playing cribbage in the cookhouse," I answered.

"Tell him to warm up the plane early in the morning. If the ice is right, we'll make a test flight first," Pat answered.

The ice was barely three inches thick as Bob and I found out, after chopping half a dozen holes, but the weather was very cold, and it would probably soon be thicker, but how fast? In the meantime, Bill and the mechanic might have an unpleasant time of it.

After lunch Pat appeared, all dressed up for flying.

"I'll taxi around a bit. See how the ice behaves."

It looked all right, but when he turned the plane into the wind and gunned his motor, it seemed as if waves appeared, or was that our imagination? After a very short run, however, he was airborne, and our birdman proceeded to enjoy himself in his element, happy to have his wings again. An anxious group was watching when finally, he went into a long smooth glide, and gently, without the slightest bump, put the plane onto the ice again.

"Son of a gun! That guy can fly!" Somebody summed up our feelings.

"Sure, is a honey on skis!" was Pat's verdict. "Handles like a charm. We'll try to go tomorrow morning. Can't wait any longer. Reg, have her all set to go as soon as we have full daylight."

"You're the boss."

While we were so busy getting everything ready for the winter operations, it was repeatedly said, "As soon as the camp is ready, and the drilling is under way, we'll have a quiet time." We fervently hoped that prophesy would not come to naught.

Pat and Reg went off on their mercy flight next morning. The ice was barely four inches thick. Waiting for their return, time seemed long, longest surely for Maud. All of us kept our ears cocked for the familiar sound of Pat's motor, in the stillness of the white wilderness.

Finally, someone shouted, "Here they come!" The news electrified the camp. Everybody ran toward the landing site. A perfect landing. Pat had found them all right, as predicted, on Archibald Lake, with little gas left. As Bill said, "Couldn't find a hole in the fog to sneak through to Beaverlodge. Tried for hours. Then we landed on this here lake. The morning after we were frozen in."

A few days later, on our lake the ice was over five inches; our two aviators decided to try and get Bill's machine out. Two small drums of gas and the two grease monkeys made the first load. I had hopes of going along on the second trip, but Pat didn't think much of that idea.

"You'd better stay here and look after things. You are in charge while I am away, and I want you to keep busy on that operating table for the birth of the baby. How is it coming?"

"OK, I'll give in, but do me a favor and take my camera. I'd like some pictures, especially of the take-off from the ice on pontoons. Pretty risky, isn't it?"

"Yea, he's taking a chance all right. The ice is rough, and he's got a load. The take-off is not the worst though, he's got to land on open water at Beaverlodge. In the meantime, he wouldn't know what happened to his floats. Well, here goes nothing. Keep your noses dry."

Again, we waited and wondered for hours consisting of many long minutes, until again the sound of an approaching plane grew from the south.

"Bill made it all right. You should have seen the damned old crate tear over the rough ice. Here is your camera, I got you some good shots, I hope." That was Pat's laconic report of the risky operation, another outstanding feat in the story of the Northern bush fliers.

I went back to my job of fashioning an operating table, with lots of patience and few tools, out of the trunks of some fire-killed jackpines. Nobody could say we didn't properly prepare for the expected baby. Maud would have somewhere proper to deliver the new one. Pat and I pooled the few remembered details of such apparatus. Concerning the construction itself, I had to go back to the methods of our forefathers. No bolts or hinges were available to us, my only tools: hand saw, hammer, brace and bit, an axe, and a chisel. Pegged construction was the answer then. There was something massive and monumental about the thing-a-ma-jig, by the time it was complete with a three-sectional top, and even stirrups. It took two husky men to move it. Doc, whom Pat had brought into camp was quite impressed by it. So far so good.

"Now we have to build a crib," said Pat, "no rest for the wicked."

"You can say that again. Who talked about a quiet winter in the bush?"

Almost every day some new problem cropped up that taxed our ingenuity. My hobby of carpentry, pursued since early youth, and the knowledge gained by working with Norwegian and Finnish masters of the art on building projects in the Laurentians found a full and satisfying application. Core boxes were needed now, and for our comfort, Laurentian-type chairs, tables, cupboards, and bookshelves.

The cold grew more intense, part of our crew was constantly busy supplying the camp with sufficient firewood, for the kitchen stove and airtight heaters. At night sounds like thunder caused by the heaving ice on the lakes, would echo back and forth, telling us that "she is a cold one" again. How cold was it going to get? Repeatedly our thermometer had registered 35-40° below zero. Winter had just begun.

At Pat's cabin, out of consideration for the "weaker sex" an annex had been built to house a chemical toilet, which had been ordered with great foresight. One morning Pat appeared at our tent, with a crestfallen expression on his face, "You'll never guess, but we are in serious trouble."

"Let's have it."

"You know that chemical toilet?" Pat asked.

"Yea, I bet it's frozen solid. It was a dammed cold night."

"Yep. Want to give me a hand?"

Discussing ways and means on our way over to the cabin, the solution dawned on us. A long, stout stick pushed through the handle of the container in question, Pat and I each grabbing one end, in our free hands carrying a gasoline blow torch, we made our way into the bush, a respectable distance away from camp.

"We must look like Chinese farmers. Ever read about them, Pat? First thing in the morning they carry the pails to their fields, completing the cycle, as it were."

"Chinese farmers be damned. I just thought what my old man would say if he saw me now."

"I like the Chinese idea. We should have a garden patch, too."

"Do you think this is far enough?" We were puffing a little by now.

"It'll be alright if the wind doesn't shift. Let's turn her over."

Then we let the hardly visible flames of our torches play against the sides of the can. After a considerable while there was a soughing sound… and we could return to camp.

Pat and I made a few more trips like that, but finally decided it was better to build a little outhouse in the bush and put a stove into it. Not a perfect solution, one side of you would be roasting, and the opposite end would be facing maybe 30-40° below. However, that was the best we could do.

So, the days went by. Flying was resumed, freeze up was over. Mail, papers, and magazines reached us again, and Pat flew the core samples to B&M for assaying. At first the results were not encouraging, a lot of traces of gold, a lot of nils, although the drill cores showed more mineral than we had ever seen in our trenches. One cold night, a bunch of us were sitting in our tent, some reading, some writing, a quartet playing cards. Bob came bursting in, "Caribou on the lake. Caribou. Thousands of them!"

He grabbed his rifle, rushed out again. Everybody else who had a shooting iron did likewise. Fresh meat was in the offing. Bully beef and canned sausages had become rather monotonous, in spite of Dan's efforts to bring variety to our fare.

The camp was deserted in no time. Only Reg, who had been dozing throughout the commotion in his own workshop, and myself, were left.

"We'd better stay here," I suggested to him. "The camp shouldn't be left alone; we have to watch the fires."

Then we heard the first shots, from the north, and soon it sounded like a small battle. Before the sound died down, Bob came back into camp. He, a true hunter, and woodsman, was disgusted. "The goddamned nitwits, shooting any which way, and never hitting. They chased the herd to hell and breakfast!"

"Did you get any, Bob?"

"Got one, not far from here. Can't go now to get it. It's worth your life to be on that lake, with those maniacs on the loose! Damn them anyway!"

"Pat will have to read the riot act to them. We'll speak to him."

So, we waited till the mad shooting died down and a few more fellows drifted back into camp. Reg and I then took a toboggan to bring in some caribou, if we could find any. We located the one Bob had felled with a clean shot. Thinking there might be more, we just got on to the open lake, when bullets began to whistle around our ears. Down we went flat on the ice.

"The bloody fools take us for caribou," grunted Reg, and together we vented our feelings in unmistakable language, lying on the ice and getting colder every minute. A few more shots, then silence, and hastily we scrambled into the shelter of the bush.

"You've got to put a stop to that nonsense," we told Pat that night, and although at first reluctant to say anything to those keen hunters, he said, "It will wear off. It's just hunting fever that got them."

But he finally laid down the law, "From now on, no more than three men at a time will go hunting. They will stay together so that they know at what they are shooting. Any stupid fool who shoots in the direction of the camp or drill set-up can consider himself fired. We want meat, not manslaughter. When you shoot, shoot to kill. If you wound a caribou, follow it and put it out of its misery."

When daylight came the following day, a whole bunch of fellows were raring to go hunting again. It seemed as if the warning of the night before was forgotten.

"You heard what I said last night," roared Pat, "you, and you and you," pointing at three fellows, "can go today. You others better get to work, what in hell do you think this is, a hunting lodge? You three guys stay on the east side of the lake and shoot only to the north and east!"

Off they went. Then Pat turned to me, "Let's go over to the drill. Take your gun along. Maybe we'll see some caribou on the west side."

At the drill we found everything going alright, the old Ford engine roaring, drill rod vibrating in the chuck. The sludge that came bubbling out of the drill hole looked blackish. "Sulfides! Must be in the vein."

"I wonder if we'll get some values this time."

Plowing our way through powdery snow in a north-easterly direction we hit the wide part of our lake. Before us in the bright sunlight lay the Green Islands, thickly wooded, unscorched by the fire that had devoured the bush on the west side of our lake a few years before our coming. Facing east, we sat near the shoreline; maybe some caribou would show up, maybe not. We weren't greatly concerned about that. It felt good just to sit there in the sun, looking far over the lake into the white shining country.

"The friendly North," mused Pat. "Now I understand it, how about you?"

"I thought about something similar just now. 'Peace, blessed peace,' a good friend of mine once said that looking over the blue hills of the Laurentians from his cabin high on a hillside. That moment often has come back to my mind. He said it, like the prayer of a man who has tried very hard, and in spite of giving his best hadn't reached his goal. I know his wish was never fulfilled. Here he might have found the peace his soul was yearning for."

A couple of caribou showed up on the opposite shore. A volley of shots rudely shattered the silence. Frightened animals spilled onto the lake, fleeing in all directions. Again, and again shots rang out. "The stupid bastards are shooting wild again," swore Pat. "Wait till I get hold of them. What did you say, 'Peace'?"

"Yea, and only man is vile."

Suddenly the firing stopped, and to our ears came a cry like somebody in pain. "What the hell was that? Do wounded caribou cry??" Pat wanted to know.

"Never heard of it. Probably one of the guys. Let's go and see."

We heard no more shots so we started across the lake. Halfway across Bob came running to join us. "Somebody is hurt over there," he shouted. "I've heard that sound before."

Two fellows came slowly out of the bush, one supporting the other. "Fred got it in the leg." He looked somewhat green in the face. Just above the knee his pantleg was torn.

"Can't be very bad. You can still walk, can't you?" Pat said scornfully. "You better try to get to camp quickly. Don't make a mistake though, I didn't hire the nurse for you guys."

It turned out to be only a small wound, which was painful enough, I suppose. The bullet had drilled a clean hole about three inches long through the flesh, just below the skin. Pat was all set for cutting it open immediately, but the patient wouldn't consent.

Thankfully, Nursey had been brought to camp for the expected blessed event. The presence of Nursey made a silently suffering hero out of the amateur hunter, who only too willingly gave himself into her care. It was never determined who fired the shot, but at least there was no more wild shooting.

Instead of bully beef and sausages there was roast caribou and steaks on our menu, a delightful change, but after a while that began to pall too. Dan, the cook, doing an otherwise clean and efficient job, wasn't very strong on variety. Day after day stretched into weeks of tomato soup and caribou steaks. Somebody suggested Cream of Tomato Soup, made with real milk. Well, that's what we got from then on. Another bright idea popped out of somebody. Meat pie, and meat pie it was from then on. The crust wasn't bad.

No wonder the whole camp fell in love with Mrs. Mack, wife of the diamond drill foreman, when she came into camp just before Christmas. A lovable, friendly, patient woman and a wonderful cook, a mother to the whole bunch, and good company for Maud, she took little hard-pressed Dan into her special care, and a good pupil in the art of cooking he turned out to be. However, before those meals, the like of which we had only dreamed and talked about, became reality, two other events kept us in an uproar.

CHAPTER IX
GOLDRUSH

UP TO THIS time, whenever we heard a plane nearby, we were fairly sure that it was approaching our camp, because there was, to our knowledge, no other outfit in our vicinity. There came a day, however, when again and again planes passed, streaking across the sky to the west and north of us.

"Seems somebody made a strike," said Pat. "I'd better go to the village of Goldfields to try and find out what's going on over there." When he returned, he brought the news, "A fellow by the name of Jim Randell is supposed to have struck it rich north of Alder Lake. They talk about $5000-$6000 to the ton. That's a lot of B.S., of course. But they are all rushing in and staking like mad."

"Which way are they going?" I asked, pulling out my well-worn map of the area.

"Somebody said the lead runs north-east."

"Well, if that is true, we'd better get busy. Look here, on the map. They will get to the north end of our lake, and we wanted that ground for ourselves."

Early the following morning we started. Six of us tied on our snowshoes long before daylight. It was bitterly cold, 50° below, stars were blazing in the sky with the bluish fire of diamonds. There was no talk as we trudged northward, tuques pulled over our foreheads, and the parka hoods tightened. Just the sound of our breathing, the swish of our snowshoes, and the creaking of the rawhide web.

At the head of the lake, we split into three groups of two to run the three lines with the No.1 posts first. We would meet at the end of the middle line. The six of us had staked many claims before. To run lines for 18 claims didn't seem like a big job. But the going was tough through two feet of powdery snow. Even on snowshoes we went right to the bottom, got caught on fallen trees that littered the ground. Deep draws had to be traversed, and steep cliffs to be scaled. But all went well, and we met in good spirits. Bob, the experienced woodsman, had a roaring fire going when Ronny and I showed up, and we enjoyed our steaming mugs of java, roasted canned sausages and hunks of bread.

"Where do we go from here?" someone asked.

"If you fellows are game, we'll try to finish the western borderline of this group. That means quite a bit of hiking, but the claim lines will be safe after that. We might run into Jim Randell's camp that way. That ought to be interesting," I suggested as a plan.

So, it was agreed, and refreshed, we set out once more on our arduous journey. However, by the time the job was finished, daylight had gone, weariness was in our limbs, and a long way was left to go home…

Our western-most claim line ran along a deep draw running north and south. It seemed the easiest way to travel. So, we trudged along in single file, hoping to find an easy way eastward towards our lake as we went along. Instead, we ran into the roughest terrain yet encountered. In places we had to crawl up the hillsides on all fours, then, sliding down on the other side our snowshoes would catch on a fallen tree or other obstructions. Sliding and tumbling we'd get to the bottom of a draw only to start it all over again, cussing and swearing, damning the whole North country to hell. A terrible tiredness crept into our minds and bodies. Our tempers were frayed.

After one more slide to the level of a small lake, Art reported he had lost his axe. "Damn glad to be rid of it too, the bloody thing weighed a ton."

Another fellow, usually very quiet and sensible, grabbed his axe by the end of the handle, swung it around his head a few times, then let it go, "There goes another one. It weighed TWO tons."

"You damn fools, you may need them before you get out of this wilderness."

"So, even you don't know whether we are going the right way! Wouldn't it be better to build a lean-to and stay here?"

"Ah, pipe down if you want to stay here overnight, it's OK with me! What do you say, Hermann?" came from Bob.

"Let the majority decide. I figure we are just about on a level with the Green Islands. About three claim lengths to the west."

"That's about my idea," said Bob. "Now make up your minds, you guys!"

We went on all right, the country flattened out somewhat, the moon came to light up our path. Suddenly we ran into snowshoe tracks running in south-easterly direction.

"Maybe those are ours," somebody said, "Maybe we are just running in a circle."

"Ah, nuts, none of us has bearpaws. Let's see where this guy went. He didn't pass very long ago. Figure he must be heading toward his camp." answered Bob. And so, a few minutes later we came to a small lake, and on the opposite shore a light was glinting. We made for it.

"Hello this camp!" We hollered.

A tent flap was pushed open, a head appeared addressing our bunch. "What in hell are you guys doing here? Did you bring your own grub??"

We were a little dumbfounded. That was a hell of a welcome! With his first words that fellow had committed one of the worst sins against the code of the North. Regardless of who comes or how little you may have; you share it with your guest. That is Northern hospitality. In the state we were in, tired out and weary, this fellow's attitude didn't sit very well with us. However, his partner had more sense; opening the tent flap wide he said, "Come on in you fellows. Are you from Pat's camp? I am Jim Randell."

"Yea, we are with Pat. We heard about you fellows being over here somewhere. We came in a round-about way though. How far to our camp?" I asked.

"About two miles, I figure. Easy going though. You will get onto your lake soon. I'll make you a cup of tea. Make yourselves at home."

That guy knew better than his surly partner, and we were glad to accept his offer. Not far from where I sat, I saw a few hunks of rock lying. Picking one up, I noticed it was a finely grained grayish-black specimen, which we had frequently encountered on our own claims without paying much attention to it.

"This what you are getting your values in, Jim?" I turned to our host.

"Yea, but we have some other stuff that kicks even better."

"So, you really got it?"

"Yes, biggest thing I ever struck. Six thousand dollars to the ton. We got it all staked though."

Well, either we were damn fools, or he was the biggest liar alive. A good smooth talker in any case and a good host. After the tea and a smoke, we started on the last lap of our journey, once more in good spirits. One more range to cross, and before us, bathed in moonlight lay the Green Islands and beyond them showed faintly the lights from our camp. All the weariness was forgotten, and suddenly there came alive again the main topic of the last weeks when Art said, "Bet you anything the baby arrived while we were slugging through the goddamned bush!"

Everybody had been deeply interested in the impending arrival of a new citizen, every man who came off his shift on the drill, or from his work in the bush, always asked the same first question on returning to camp. "Has the baby arrived yet?"

There was a great deal of tenderness and concern hidden under the rough outside and the rougher language of our crew. A few days ago, Pat had brought the Doc from Goldfields. The time must be near.

Bob and I made for Pat's cabin to find out. Pat must have read the question on my face when I came in. "No baby yet, Hermann! How'd you make out? We thought you guys were lost."

"Some of our gang thought so, too. We got the claims secured. Dropped in at Jim Randell's camp."

"You mean to say you staked all the claims on a day like this? Holy cats what did the other guys have to say?"

"About the claims, we got to run some cross-lines yet. It's hellish country. Posts are all in, however. Had a strange welcome at Jim's camp. Guy by the name of Hank Moles, that's Jim's partner, you know what he said when he saw us? 'Did you guys bring your own grub?'"

"He didn't! Holy mackerel, but I do hope that guy makes the mistake of walking into this camp sometime! Well, you fellows must be played out. Better take a load off your feet. I'll buy you a drink."

Pat's invitation sounded very enticing. There was a nice domestic scene before our eyes. Maud busily knitting a tiny garment, the others reading, soft music came over the radio.

"We'll be back for that drink after we have cleaned up a little, and got something to eat out of Dan. Even meat pie sounds good tonight."

"OK We'll see you later."

Dan, willing and obliging as always, got busy while we cleaned up, and boy, did we ever need a good wash. Clean, full of, yes, meat pie, tired, yes, but somehow feeling gloriously alive, we went to get our well-earned drink from Pat. We gave him a detailed report of the day's work, and of what we had seen on our journey.

"When do you want to finish the claims?"

"Tomorrow. We thought you might run us up to the end of the lake. That would save us a lot of hiking, give us a little more time to look around, and maybe, pick up some samples."

"I'll do that. Reg says the motor is OK now; it was missing on one cylinder. That's why I couldn't take you today."

Long before the sun had started on her short winter-day's cruise, we were on our way next morning. Four men squeezed into the cabin, two men sitting on the lower wings, holding on for dear life, as Pat taxied us to the north end.

Shortly after we began cutting our crosslines. Ronny suddenly called out: "Look over there. A tent!"

Sure enough. Well hidden in the undergrowth, there it was.

"Do you think Pete Lauder is still around here?"

"Let's go and see. Who else could it be?"

Just before the snow came, Ronny and I, on one of our prospecting trips had run into the old-timer, a big, barrel-chested fellow, around 60 years of age he looked, a bushy, greying beard hiding his features, but not a pair of blazing blue, friendly eyes. He had taken the wind out of us, when he offered to take us to his showing which he had described as a massive lead of copper pyrite and galena in conglomeration. "It's not very far from here," he had said, after we dropped in at his camp, following a freshly blazed line. We called him "The bull of the woods" after that trip.

Uphill and down he travelled with terrific, silent speed, his enormous chest hardly heaving, while we young fellows had to labour to keep up with him. At that time, he had told us that he would soon go to Goldfields and accepted our invitation to stay overnight at our camp. In the rush of events, we had almost forgotten the old fellow.

Now we approached the tent with a sense of foreboding. When we opened the tent flaps, we saw him lying in his eiderdown on a bed of boughs. His eyes were feverish, and in a croaking voice he greeted us, "Hello there. I thought you fellows might drop in some time."

"Hello, Pete. What's the matter? How long have you been like this?"

"Took a chill about a week ago. I'll be all right in no time. Wouldn't have happened if I had had some Rum left. Always take a tablespoonful when I feel chilly. Best medicine a man can carry with him in the bush."

We looked around his tent. There wasn't much grub around either.

"What are you living on, Pete. There isn't much one can see around here."

"I got enough to eat," he bravely lied, "but I ran out of tobacco. That's worse."

"OK, Pete. I can give you enough to last you till we come back tomorrow. We are finishing our claims and Pat is going to pick us up tonight. We'll tell him about you. In any case Ronny and I will be back with some grub and medicine tomorrow for sure."

"You shouldn't worry about me," he protested, but not too strongly, and there was a sort of wondering look in his eyes, as he looked at Ronny and me.

Was it because he wasn't used to being worried about? Was it because sometime in the past somebody hadn't cared enough? Was that why he now found himself all alone and sick in the Northern bush? Thoughts like that went through my head, when we left Pete, after again assuring him that we would be back the following day. Ronny's thoughts must have run along the same lines.

"I wonder," he said, "what makes a fellow go off by himself like that?"

"Well, we can only guess. Per is another case. They will never tell, I suppose. My guess would be an unlucky love affair."

"Too bad," said Ronny. "He seems such a nice chap. We must make sure not to forget him. But now we better remember that we have some lines to blaze." Well said, I thought.

That kept us hopping for the remaining daylight hours until we joined up with the rest of the gang.

Hardly had we met when Pat came roaring down the lake to taxi us home.

"No news yet, Pat?"

"No news, but it won't be long," Doc says. "Kripes, I wish it were over."

At camp we related our encounter with old Pete and, we won Pat's full support for our plan to go back with some supplies the following day.

"I haven't got any rum left, but you can take a little scotch to cheer him up. Tell him he's welcome here any time."

To our great surprise we found our new friend up and around, when we reached his tent the next morning.

"Did me a lot of good to see you boys yesterday. Feel much better."

We unpacked the grub and tobacco we had brought. And last but not least the scotch. Pete protested vigorously, but through his protestations there shone genuine happiness, and that made us feel good too.

"I'll make us some tea."

"No hurry, Pete, take it easy, and take the medicine first. We'll stay for a while. We want you to come to our camp if you care to. Stay as long as you like," we told him when the tea was brewed.

Again, there was that wondering look in his eyes. He didn't say "yes" right away. We didn't expect it, as we knew he would have to get himself used to that idea first. So, we sat and talked, drank lots of strong tea and smoked. When we left, we just told him once more, "We will be seeing you at the camp then. So long."

"So long fellows. Thanks for everything."

In camp the expectancy had reached fever-pitch by now, the arrival of the baby wasn't just a private affair of Maud's and Pat's. It seemed to be everybody's greatest interest, and main topic of conversation. Finally, after a few more long days, one dark evening, Pat burst into our bunkhouse, storm lantern in one hand, box of cigars under his arm, a bottle in the other hand.

"Have a cigar boys." The new boss had arrived! "Have a drink."

Cigars were passed, the bottle made the rounds, everybody pumped the proud father's arm, and there was great shouting and rejoicing, loud enough to tell the rest of the gang in the cookhouse and the other bunkhouse of the good news.

"How is Maud?" I finally managed to inquire.

"Oh, fine, just fine. Come on over and see mother and son."

Everything was just fine indeed. Maud smiling and composed, looking grand, as if nothing at all had happened, but there was the new little man too. We met Pat, in his smooth-skinned perfection. Everybody was very happy.

CHAPTER X
DEEP WINTER

CHRISTMAS WAS APPROACHING. Turkeys had been ordered from the outside. Little Dan, the cook, was running around with a deep frown on his face, until, just in time, Mrs. Mack arrived to take charge of the preparations for our feast, and lovely, mouth-watering smells filled the cookhouse, as Christmas pudding, cakes and pies were produced in profusion, each one a succulent masterpiece.

A long festive table laden with mountains of heavenly food, a tall Christmas tree hung with colorful camp-made decorations, the gang with scrubbed and shining faces, among them Per and Old Pete, Pat and Maud and Nursey, and the little newcomer Little Pat, well wrapped, brought over in a tin washtub. Mrs. Mack and Dan scurrying around adding final touches, and so we sat down to our first Christmas dinner in the bush, to enjoy the blessings. Short, heartfelt speeches were made, Christmas carols were sung, and happiness filled the hearts of all present, including the two hermits, Per and Pete.

When the much-enjoyed party broke up in the small hours of the morning it was unanimously decided, "We'll have another such party at New Year's."

Again, the cookhouse was filled with great doings, as the last day of the year approached. Since Mrs. Mack had lent a guiding hand and taught our Dan many new tricks, our meals had been transformed into something, not only sustaining, but utterly enjoyable, and to this day I recall that New Year's dinner in the bush as one of the most memorable meals in my life: Caribou roast done in a manner to make it an epicurean delight, golden crusted pies in great variety, luscious cakes covered with rich icing, honey-dipped doughnuts, cookies and camp-baked bread. Art who had often talked about the "thick steak smothered in mushrooms", whenever the yearning for the bright lights struck him, made the most impressive comment, "If things go on like this, this here little Art will stay in the bush forever."

Pat had seen to it that there would be plenty of good hard stuff too, to toast the New Year. Before the party got under way Pat asked me to pick two fellows, who should act as policemen for the camp, to see that everybody got safely into his bunk, once the party was over. Trusted Bob and Punk, a little man who had joined our gang in the Fall, one of our most cheerful and best workers, solemnly promised to stay away from the firewater until the merrymaking was over.

Again, we sat down to a sumptuous meal with all the trimmings, to the accompaniment of toasts and songs, mouth organ and accordion, many an elbow was bent, many a glass raised on high. And so, we said goodbye to the old year, which had brought us together in the Northern bush, and with hopes strong in our hearts we greeted the future. When the revelry was over, Bob and Punk, true to their pledge, guided the revelers safely to their quarters, and silence fell over our camp.

Don and I sat up for a while, talking and reminiscing. After an hour or so we began wondering why Bob and Punk, who slept in our tent, hadn't shown up. So, I put on my parka once more to check up.

The first place I headed for was the cookhouse, and no farther did I have to go. In the middle of the floor our two policemen were sitting, hugging a fair-sized keg with their legs. Unknown to all, Dan had started to make some wine from dried fruit. How, I don't know, but these two had found it. Each one had a soup ladle. They were very polite to each other. Holding out a brimful ladle Punk would say, "Have a drink, Bob. You are a fine fellow."

Then it would be Bob's turn. "Have a drink yourself, Punk, you are OK."

"Have another one, Bob. It's good stuff."

"Your turn now, Punk."

"Duty before pleasure, I always say. Everybody safely in bed, now we celebrate a little! Damn good policemen! Have another one Bob."

Their tongues were getting thicker with every ladle full. They hadn't noticed me coming in at all, so I finally asked them, "You guys going to finish it in one session?"

Immediately both dipped in their ladles once more. "Have a drink it's New Year's!"

At that moment Dan, who had his bunk in the cookhouse stuck his nose out of the sleeping bag, "For crissake get those nuts out of here. I'd like to sleep. Too bad they found the stuff; it's not even done yet."

It was a yeasty but potent concoction, alright. I had to try it; Bob and Punk were very insistent.

"You'll all kill yourselves. Get the hell out of here," Dan roared once more, but my powers of persuasion were not as strong as Dan's dynamite, and our two friends just kept on ladling. It was best to leave them to their fate.

Quite a while after Don and I had hit the hay to sink into dreamless slumber, the door to our bunk house flew open with a crash, and with his legs going independently in all directions, little Punk, with husky Bob draped over his shoulders, came staggering in. He made for Bob's bunk, and dumped his burden, all the while talking to his unconscious chum, "Don't worry, Bob, you are all right. Don't you worry, Punk takes you home. Don't worry and go to sleep. Punk takes care of you. Yessir."

And then the faithful soul staggered to his own bunk.

CHAPTER XI
REALLY COLD

EXTREMELY COLD WEATHER had been with us for a long time. One morning looking at the thermometer, we couldn't even see the alcohol column, it had retreated below the minus 60° mark. We could hardly believe it, but a little later, when we went to see Mack in his cabin, he had proof for a greater cold than any of us had ever experienced. He demonstrated to us that the acid mixture he was using to burn off the metal of the used diamond bits, wouldn't come out of the container, it was semi-solid.

A few days later a change set in. There was a different feel to the atmosphere. At first the cold seemed more penetrating, although the thermometer began climbing. Then it began to feel warmer, in spite of the temperature being still way below zero. That afternoon near sunset time, on our way over to the drill set-up, Pat and I saw darkish clouds with peculiar yellow-orange linings drifting before a western breeze.

"Looks like a Chinook sky," said Pat. "But I don't think a Chinook will reach this far." We all admired the straight line of clouds far above our heads.

However, the next morning there was no doubt, it was indeed the warm wind from the Pacific that made the water drip from the roofs of our camp. For a few days. it seemed like Spring, but then a new snowfall heralded more cold weather.

It was about this time when once again an old problem cropped up, which would not be worth mentioning, hadn't it been for the spectacular solution, which was retold again and again around our camp.

Our heated outhouse had proved a great success, so much so that we were now facing the fact that the holding capacity urgently had to be increased. Punk was put in charge of the operation. Together we looked at the proposition and decided that we would simply dig down behind the present, insufficient excavation, and then move the little house back over it. Easily said, but the solidly frozen ground made the picks glance off. However, it had to be done. So, I left the little gang, and returned to my occupation of logging the drill core and preparing the samples.

After a good while I decided to have another look to see what progress was being made. When I started to go up the well-trodden path, Punk came running. Not seeing me, he went behind a tree. At that moment something happened that seemed to throw him into great excitement. One of the girls had stepped out of Pat's cabin to follow another well-worn path, leading to the same much-frequented location. Punk jumped up, waved his arms frantically and hollered, "Go back, go back" and not finding the right word, loud and clear it rang out: "SSSSShiit." A deafening roar, our little outhouse turned a somersault, lifted up by a mighty, brownish eruption. Everybody in camp came running.

"What happened?" Pat asked in an ominously quiet manner.

Punk stammered explanations, "Couldn't dig. We used dynamite!"

Then there was another, verbal, explosion, "You crazy, goddamned, stupid fools, you could have blown the whole camp to kingdom come!" addressed at the badly shaken Punk and his crew. "Get busy, you guys and clean up the mess. Anybody using powder again near this camp is fired, and he can walk out, too!"

We never found out how many sticks of dynamite the boys had used, it must have been a goodly number. Where the little house had stood was just one big gaping hole.

"Good enough for two winters," somebody said.

"Ever see freckled snow?" another remarked.

At least once a week Pat would fly to Goldfields to get our core samples to the assayer, and to collect our mail. One day, Mack, the drill foreman, went along. While Mack was gone Mrs. Mack, the good and admired friend of all took some time away from the kitchen to spend the day with Maud. The daylight hours were still only few. As soon as it was bright enough Pat had taken off.

"Keep an eye on the girls in the meantime, Hermann. I will be back as soon as possible."

But early twilight came, and no Pat, no sound of an approaching plane. The ladies were getting nervous, and so were we, but we tried not to show it. Darkness enveloped the camp, not even the moon appeared, to give us hope that our plane might still return.

"What do you think has happened?" Again, and again the question was asked. I tried desperately to find some plausible and at the same time un-alarming explanation.

"Maybe they had trouble starting the motor again. It's cold enough tonight..."

That seemed the least alarming answer.

"What if the motor failed before they got there or after they took off again?"

"Pat can always get the plane down; he follows the lakes."

Worry and anxiety crept into our minds and reasoning didn't help much anymore. The worst thing was, we couldn't do anything to end the uncertainty. Our radio was tuned to the B&M station, but as often before, all we heard was crackling static. So, the long, dark hours trickled away minute by minute. Around 10 o'clock there was a knock at the door of Pat's cabin, where I shared the anxious vigil with our ladies. I went to the door. Outside stood Bob, all dressed up, snowshoes on his feet, packsack on his back.

"I'm going to Goldfields, Hermann, to find out. Tell the girls I am going, that will help a little. Better than doing nothing."

"It's 25 miles on foot at least, Bob. Sure, you can make it?"

"Sure, sure!"

"Better take somebody along, Bob. It's no good travelling alone."

"No, I am going alone. Make better time."

"Got any grub?"

"Yea, I am going to make myself some tea at Per's cabin. That's about halfway. So long."

"So long, Bob. Good luck!"

Good faithful Bob. A better friend no man could wish for. His going had a reassuring effect, for a while, at least. "But how long will it take him to get there?" someone asked.

"About six to seven hours," I replied, "Bob can travel! At least he can start some action if he doesn't find them at Goldfields. But I think he will."

Doubts were in my mind too, but I believe I managed to put on a good show of optimism. I had full confidence in Pat, based on our past experience, that he could pull himself out of any tight squeeze. Still, the uncertainty was hard to bear. Suddenly a thought flashed through my mind, "What day is this?" We never paid much attention to that question under normal circumstances, now it was highly important.

"Saturday."

"Saturday? That's great. At 12 o'clock we'll get a message through *Hello, the North*", I was positive.

"How could that happen?"

"Pat will send a message through B&M or the RCMP to CJAD in Edmonton, and we will hear on that station. Let's tune in on it now."

CJAD came in clear and strong. We were not always that lucky. Like every Northerner who had a radio, every Saturday at midnight we would listen to Edmonton to hear "Greetings and messages from the outside world sent to the far-flung outposts of the North." I took the strong signal as a good sign. I prayed hard, and so, I believe, did all our gang.

Eleven o'clock. One more hour to go. Would it ever end? An hour can be an awfully long time. This one definitely was a piece of eternity. Finally: Midnight! All of us moved closer to the radio. Then it came, "CJAD Edmonton. Hello the North! As usual every Saturday at Midnight CJAD sends its program to its many friends down North. Before we get our program under way, here is a special message to Pat's camp at Neely Lake. The message reads: "Held up at Goldfields due to motor trouble. Everybody is fine. Return as soon as plane OK. Greetings!"

Several times the announcer repeated those most welcome words. A wonderful feeling of relief came over us. We all swallowed hard for a few minutes, and in our hearts, we silently thanked the Lord. Now we knew, and could go to sleep, unworried, but our friend Bob was still on his lonely trek through the wilderness, alone with only the stars and the aurora borealis lighting his way. My thoughts tried to follow him, until I fell asleep.

The next morning, I had to face another problem. During the night, the diamond drill had been pushed through the lead and had not found anything worthwhile in this location. A new set-up was indicated. One of the drillers, let's call him Mort, was in charge of the drill crew during Mack's absence. He came to me to find out to which place he should move the drill. I gave him a line-up which necessitated a set-up on the ice on account of an indentation in the shoreline. We were anxious to push down a hole on that particular spot, because a very noticeable, mineralized quartz outcropping adjoined our vein there, maybe it was even a part of it.

Mort and his crew dismantled the tripod, drill, and waterline, but when all was ready for moving, he refused to go to the spot I had lined up. Nothing would budge him. He wanted a set-up on land. I knew it was easier to collar a hole on solid rock, however we had drilled from the ice before, there was nothing unusual about it. I was in charge of the camp, there wasn't any reason that I could see, to argue at length about Mr. Mort's likes and dislikes.

"Are you going to drill where we want the hole or not?" I asked him.

"I am not setting up on the ice," he replied, "I am setting up on the shoreline."

"You are not getting any depth that way, it's no good, and you know it."

The other drillers had joined the confab. I turned to them, "You fellows of the same opinion as this guy?"

They didn't seem to feel too comfortable about it, but thought they had to stick to their temporary foreman.

"OK. You fellows are trying to pull a strike. You can have it. There won't be any drilling. You can sit on your asses till Pat and Mack come back!"

Our own fellows, some of whom worked as helpers on the drill found enough work to keep themselves busy, but there was tension in the air, even the ladies grew concerned as the drillers sat out the long hours in their bunkhouse. I went to see them, "You guys changed your minds by any chance?"

Some of them, I felt sure, didn't feel any too good about the whole thing, but Mort still had them under his influence.

"Just thought I'd give you guys one more chance. That's all. After this, tell your story to Pat and Mack."

At the end of the third day of the strike, a two-man delegation came to me. They wanted to resume work, Mort, or no Mort.

"Too bad, fellows, but it's a little late now. You let Mort bamboozle you into this. You'll just have to sit it out. Sorry."

I counted on Pat's return the following day, but two more days passed before we heard the familiar sound of his plane. First thing he asked, after pulling himself out of the cockpit, "Why isn't the drill working, Hermann?"

"Drillers are on strike. Wouldn't drill from the ice."

"Well, of all the dirty, lousy, stinking sons of bitches. Why didn't you fire them?"

"I saved that for you. I'd like to say that only one guy is to blame for this. Don't be too hard on the others. They suffered."

"Suffered? My foot! I'll teach them a lesson."

Mack, and Bob, who had reached Goldfields the morning after he left camp, had joined us. Mack was furious but didn't say much. Pat didn't mince words; the wrath of the Irish was aroused.

In spite of all this there was a joyful reunion, between husbands and wives, father, and son, with our own crew and Pat, Mack, and Bob.

Later on in the evening, when Pat and I discussed the unpleasantness of the strike, Pat remarked that Mack's quietness had astonished him.

"It's a funny thing. I knew Mack before he came here. He's not the same man. I wonder…"

"Do you mean he's afraid of someone?"

"Yea, that's about it. And there's another thing. Some lowdown bastard is sending out reports from this camp. I found that out. There's nothing to hide here, I think we still have the best goddamned camp in the country. I just don't like to have a snake-in-the-grass in our midst."

"Well, Pat, things are beginning to fall together. Have you ever heard of a farmer who sends long, weekly letters to his son, or of a farmer's son writing long letters to his old man every week?"

"Who the hell does that?"

"I thought you might have noticed it. You are taking and getting the mail. Anyway, since you fired Mr. Mort, there won't be any more of it."

Mr. Mort took the next plane out, and Mack was a changed man, as if a great burden were taken off his shoulders. We never found out what was behind the affair. In the North you mind your own business, and you expect everybody to do likewise. You are close to those with whom you happen to live in a camp, because they are practically your only company, but your intimacy is not forced, there is no prying into a man's private affairs.

CHAPTER XII
VACATION TRIP

WHEN PAT AND I first set out on our trip into the, to us, unknown, it was with the understanding that we might be in the bush only for one summer's prospecting, and with this in mind I had been planning on visiting my family in Germany this year. It was 1936, and six years since I had come to Canada. To go home for Christmas seemed especially tempting. However, our good luck, the finding of a promising show and the resulting activities, camp building and diamond drilling, made me postpone my trip. If I still wanted to go to Europe and wanted to be back in Edmonton right after break-up to return to the North, this was the time.

Discussing the plan with Pat during one of our long "every evening meetings" in his cabin, it was decided that Pat would take me to Goldfields the following day. Often in the evenings, when the news came over the radio, or as the result of reading magazines and newspapers, we had gotten into lengthy discussions of world politics and lost ourselves in speculations about the eventual outcome of the generally muddled situation. The noise from the other side of the Atlantic had grown louder and louder since 1933, many conflicting reports had made it more and more difficult to form an opinion about the true state of affairs over there in 1936. Some said there was nothing to worry about, others predicted that there would be war. Now, I was going, mainly to see my folks, but at the same time I was anxious to have a good look for myself. When the other boys in camp heard about my intended trip, I was in for a good deal of good-natured ribbing.

"Don't come back without finding yourself a girl."

"Don't just get one. Get a few for us!"

"Just think of it, the lucky stiff, lolling on a deck chair, beautiful dames around, moonlight on the ocean, and us poor buggers, standing on a tripod, pulling rods when it's cold enough to freeze the nuts off a brass monkey. What a life."

Another one would say with great concern, "I read in the paper, they haven't got much to eat over there. Better take some frozen caribou along."

My preparations didn't take very long. I like to travel light. A comb and a toothbrush, and I am ready to go. In the North you have to take your bedroll though, just in case. On our way over to Goldfields we were to stop at Per's cabin on Beaverlodge Lake. I had spoken to Per for some fox skins. Morning came. Last breakfast in camp, and then, "So long and goodbye."

Into the north wind roared our plane, then it swung south. Just over Beaverlodge Lake the motor started to cough and miss. Through the speaking tube came Pat's voice, "I'd better go right through or you might miss your plane."

We got to Goldfields alright and just in good time. At the B&M camp Sparks, the radio operator, informed me that Ken would be going to Fort McMurray in about an hour. That gave us a chance to see Alex, the trader, who, at that time of year didn't have many furs, but dug up a few fox skins and an otter for me.

Pat and Reg went with me to the plane, whose motor was already ticking over.

A last handshake, a few cracks to hide emotion, I crawled on top of the load. Pat and Reg rocked the plane. Ken gave her the gun. Soon my two chums were just two little lonely black dots on the white expanse. In a few more moments I lost sight of them. But as long as I could still see them, an uncertain feeling of loneliness took hold of me. If I could just have jumped down, I would have run back.

Many a time on my trip this feeling returned. Like a powerful call it would ring, calling me back to the North, to my friends, which then seemed to be the only real things, the real home. "Heimweh?" Yes, that was it. The German word for homesickness described by feelings for the North. With a strong tail wind, we made excellent time to Fort McMurray, from there I was going to go on with Canadian Airways or Mackenzie Air Service. No plane that afternoon, so we parked ourselves at the hotel, after calling at the bank to get my money. The next morning, as soon as it was daylight, I went to the Airline offices at the Snye. No luck there, but when I got back to the Franklin Hotel, there was Ken, "I've been looking for you all over. Just got word to go to Edmonton. Come on. Let's go."

Good luck was with me.

"Damned cold morning, Hermann. The old crate is drafty."

"Do you want my caribou parka, Ken?"

"I would like it, but how are you going to keep warm?"

"That's easy. I'll just crawl into my bedroll."

So, we traded. Snug in my sleeping bag I fell asleep. When I looked out of the window, after what seemed only a short interval, there, below us, was the Edmonton Airport. A smooth landing, into a taxi, and then the King Edward Hotel, haven, and rendezvous of all Northerners. In a parka and mukluks, you were the King. And genial John Calhoun politician and owner of the hotel, made you feel at home.

The next thing on the program was a good long soak in a bathtub. In the meantime, a bellhop had gotten my suitcase with the city clothes out of storage. Then I headed for the barber shop, "Give me the works!"

"How are things down North?" the barber asked.

My camp style haircut and my dark skin must have told him. Down came the long locks that curled around the ears and neck, leaving a sickish pale line. Hot towels smacked into the face. Smelling like the cosmetic counter of a 5 and 10 store and properly itching from the small cuttings that every barber insists on brushing down your neck, I headed for the steamship office.

"One round trip to Hamburg, Germany, if possible, on the *S.S. Deutschland*. I need two days in Montreal and one in New York." I had it all figured out. The agent, somewhat astonished, looked at the schedule.

"Why the *Deutschland*?"

"Oh, just an idea," I replied.

"Well, that works out fine. The *Deutschland* sails from New York on the 16th of February. That is, one minute past midnight of the 15th."

I was a little surprised myself. Ever since I had thought of my trip, I had had that particular boat in mind. I had worked on it while it was under construction. Those were my thoughts, and the only connection I knew. Later on, I came to think that fate might have whispered in my ear.

A few commissions for the boys in camp, and a little business with the Hudson Bay Company took up the rest of the day. Ken would return the following day. He agreed to take a few chickens back to the camp, just to say I hadn't forgotten them.

I had a good supper with a friend at the Springer Hotel, and soon it was train time. A long, long journey. Not many passengers. A few very reserved, prim, elderly ladies, looking as if they were heading for Toronto, and a bunch of Legionnaires fighting the last war over and over again in the smoker. At long last I arrived in Montreal.

All across Canada it had been very cold. 35° below. I didn't mind. But when I got off the train in Montreal, where it was 10° above, I began to shiver. A fog was hanging over the city, damp and penetrating. Give me the North!

I rendezvoused with an old friend, having supper at the Queen's with Chris, cherished friend of Laurentian days. We shared good food, dark wine, white tablecloth, gleaming silver, soft lights, and soft music. Life was good.

I embarked on the over-night train to New York, which was digging itself out of the previous day's blizzard. That trip was followed by an evening spent with a dear friend of long standing, charming Dorothy, and someone I had never met: Uncle Henry. There was talk of the North, talk of home and a few Alamagoozlum cocktails which was a shaken rum beverage with pineapple garnish. The hours flew while we saw Radio City, Leslie Howard in *The Petrified Forest*, Rockettes, ballet, colour and music. How could heart and mind grasp it all?

I went down to the pier, and once more I stood on the deck of the *S. S. Deutschland*. Fifteen years had passed since we cut and drilled the plates for the superstructure at Blohm and Voss, the shipyard I worked at in Hamburg's harbour. It looked a lot different. Now the railings were lined with passengers, bantering back and forth with the friends they left behind, down on the crowded pier. I was all alone, but somehow it felt good, long leave-takings are not to my taste. Somewhat detached I watched the show. Paper streamers flew. Some shouted, some kept quiet, and some had tears in their eyes while the ship band played; Must I, Must I Leave, a very sad song… Behind me on the top of a hatch a dark-haired girl, somewhat the worse for drink and emotion was trying to sing: "Thanks a million, a million thanks to you."

It was midnight, and the booming roar of the boat's whistle, tugs answering, moorings were cast off, and stern first we glided into the Hudson River. The bow swung eastward and past the myriad lights of Manhattan, we headed for the sea.

Soon I found myself practically alone on deck. I stayed near the bow until we passed the Statue of Liberty, then wandered back to the stern to see the last lights of New York swallowed by the haze, my only company a bunch of seals, destined for some old country zoo, plaintively bellowing their homesickness. And so, to bed after a few foaming beakers of Munich beer at the bar, which was opened at last.

It had been a full day. The next morning, for a moment completely at a loss as to my whereabouts, I was awakened by the sun slanting into the porthole, and by strange unfamiliar noises. Where was I?

Oh, yes. We sailed last night. First day on the ocean. No time for sleeping any longer. I hurried on deck. A soft, warm, tangy breeze, sunshine and glass smooth sea greeted me. A perfect day it promised to be, how perfect, I didn't even dream, and yet I was so close to knowing.

A few trips around the deck. A gong is sounded. Breakfast time. A friendly Chief Steward directed me to a table, "A wonderful morning, isn't it?"

"Wonderful indeed!"

I was the first at the table, but soon I was joined by one of the ship's officers. Another gentleman and a girl appeared, and together we tried to do justice to the variety of a liner's menu, with a seaborn appetite.

And then, was fate pulling me by the ears? I looked up, and in walked my dream, a vision of loveliness, that made my heart grow big. I am afraid I must have been staring rudely at the next table, where the picture of all my hopes and desires took her seat. Strange thoughts crowded my mind. *Who is she? Where is she going? Wonder who the boy is, sitting next to her? Her brother? Could be. Would she travel far after we reached the home port, would it be possible to go and see her?* It was terribly urgent to get an answer to all these questions, and suddenly I knew. I had fallen in love. No question about that, and the other questions would have to be answered quickly!

Strolling around the deck with the travel-companions from my table only a short while later, we ran into the lucky boy who was sitting next to the girl of my dreams, her brother, as I thought. Don't ask me why! I must have become somewhat confused, at least very single-minded. Before I had a chance to get an answer to the only questions worrying me, the young chap told us that he was going to visit his grandparents on one of the Frisian Islands. I immediately tried to figure out the train and boat connections to there from my hometown. Maybe I had to row across?

Miss Constance, my right-hand neighbor at the table suggested shuffleboard. I turned to Dick, that was the "brother's" name, "Would you ask your sister to play with us?"

"My sister?"

"Yes, the young lady who sits next to you at the table!"

"Oh, you mean Miss Marquardt from Hamburg!"

"From Hamburg? You know, you have told me something very pleasant." With one stroke the situation had become much clearer. Where did I get the confidence that the final outcome would be in my favour, I don't know but the day was beautiful, the girl of my dreams, Lotti, was now my partner at shuffleboard, the world seemed full of promise. From then on, the days seemed all too short. Enjoying to the full every waking moment, we had to stretch the days into the wee hours of the morning. We enjoyed wine, good lively conversation, champagne and shipboard festivities, moonlight and starshine over the ocean. This, we thought, could go on forever. Uncle Henry had given me a bottle of Middle Hope wine with: "When you are happy, when you are together with very good friends, I want you to drink this!"

That moment came sooner than I thought. On the second day out of New York at breakfast time we found Miss Lotti Marquardt's table and chair flower bedecked, including a birthday cake, beautifully decorated. It was my dream girl's birthday. When night fell, after a sunny warm day spent on deck (we were still in the Gulf stream), we found ourselves together in the lounge, listening to the stories of our friends, and telling some myself. My story, unsurprisingly, was about The North.

I went to get my photos taken in Canada's North. In my cabin I remembered Uncle Henry's thoughtful gift, and it came to me. *This is the moment he must have meant.* With the rich, red wine glowing in our glasses, I told my friends Uncle Henry's words that accompanied the gift, and we raised our glasses to friendship and love and all that is beautiful in this world. And then, way out in the ocean, we were deep in the Northern bush. The northern lakes, our camps, Pat, Maud, Stuart, Reg, Ronnie, and Art and all the others appeared in pictures, and into my heart there stole once more, even in this happy hour, the feeling that filled me when I saw the last of my chums, two small black dots on the ice at Goldfields Bay and I knew then that the North would be forever calling me. But now I was sitting close to the one girl in the world who, I knew well, meant everything to me.

How could that work out? It was not just asking, "Will you marry me?" That was daring enough after an acquaintance barely a few days old. What would the answer be to, "Will you come down North?" Few men and very few women had gone into Canada's Northland, and most people would prefer to stay in the more settled districts of the country, closer to what may be called "civilization". Partly, no doubt this is due to the widespread ignorance concerning the character of the North, the fear of loneliness and exaggerated descriptions of the hardships to be encountered, mosquitoes and flies in the short, hot summer, and the bitter cold in the winter. Even most Canadians will think twice before they go and find out for themselves.

Now, I was going to ask a girl, born and brought up in a big city, in another country, to decide to break with all that was familiar to her, and to take the chance that the North might give her what I had found. I had to know, and when before parting for the night, we took our last walk around the deck, blame the moon, the sea or better yet, blame love. I asked the question, "Dearest, I want you to know that I love you, for always. Will you come North with me?"

Clear and unafraid came the answer, "Yes, I love you too." The boat didn't stop, the stars didn't fall, and the moon just kept on smiling, and yet to us a new life had begun that night on the Atlantic. She was ready to join in the adventure.

Colder weather greeted us as soon as we were leaving the Gulf stream. The seas became rougher, the wind howled in the riggings, rain and fog hid the vastness of the ocean from our sight, but nothing could dampen our enjoyment, now heightened by the secret of our pact. The nearer we came to our goal the faster the hours seemed to fly. Cherbourg, first stop, was reached one early morning, and when we crossed the channel to Southampton, sunshine broke through the overcast, painting the fields of England's south shore a brilliant green, dotted with dazzling white cottages. The steamer plowed past the Isle of Wight in lush spring colors. Night fell when the white Cliffs of Dover came in sight. Lights began to blink on both shores, and now that only one more day lay between us and our hometown, our thoughts began to race ahead.

I began to wonder how things would really look to me in the country I had left six years ago. Great changes had taken place, letters and news had told about them, and I had a premonition, strengthened by ominous warnings given even on the boat to be careful in my talk, especially concerning politics, that I would see and hear some things I wouldn't like. However, those gloomy thoughts were pushed into the background when we saw our folks on the pier at Cuxhaven.

Then there was only the joy of reunion with our families in our hearts, and last things said first and vice versa, we tried to tell each other everything our letters hadn't told.

A polite but puzzled Custom's officer had quite a time getting over my bag of raw furs and rock samples, until the train pulled in. The coaches, and more yet the compartments, seemed awfully tiny after Canadian and American trains. I still remembered the dilapidated state of the rolling stock when I left the old country. Now everything looked well-kept and in good repair.

"You even have leather straps on the windows again!" I exclaimed in surprise.

"Oh, yes," one of my sisters said. "Everything is very orderly, but…" She didn't finish the sentence. Later on, I became familiar with that phenomenon, the unfinished phrase, the watchful glancing around. Soon we were catching up on missed time. Those subjects were much safer for my part at least.

For the time being I decided to stick to purely personal and family matters, and there were many things to be recounted, so that the hours flew. Central Station in Hamburg. More relatives and friends on the platform.

My chosen one and I had kept our secret in spite of some innocent questions and inquisitive looks. Now, we had to part for a while, but it wouldn't be for long.

Among the friends at the station there was a classmate of mine, a young lawyer. Hardly had we taken our seats in a taxi, that was to take us home, he turned to me, "I don't want to spoil your home-coming, but things here are not what they look like on the surface. Just keep your eyes and ears open, you will find out. Don't say anything of a political nature. Be careful. They are listening everywhere."

There it was again. It might be said, "This has nothing to do with the North." But to me, it seemed even then it had a great meaning, more far-reaching perhaps than could be imagined. It somehow linked up with our long talks during the evenings in camp, when we had been trying to puzzle out the trend of events in world affairs, and more than once we had spoken of the tremendous temptation that the wide open and undeveloped spaces of Canada presented to would-be conquerors. Was somebody getting ready? In the days that followed my home coming I soon learned how a nation was being trained and geared for war. There was no escaping the impression. Wherever you went you saw uniforms, parades and drills in the streets and squares, and martial music filled the air. Whole blocks of barracks had been built near home, and from the nearby airdrome came the drone of motors day and night. Silvery flashes in the sky on clear day aroused my curiosity.

"Don't you know? Those are gliders. They take them up, several together. It's a new stunt of the Luftwaffe."

All these observations made me ask again and again, "What's it all about? Against whom are you going to make war?"

When I got an answer, which wasn't very often, it was always the same, "'We have to be prepared to fight against Russia."

By and by I gave up getting into any talks about politics as they usually ended by someone trying to tell me that all my ideas were wrong, and that the world, on the whole, had the wrong idea. Only a few elderly people were worried, they were the ones who remembered the slaughter of the last war. They remembered in World War I the senseless sacrifice of the best young men, the cries, and tears of those left behind.

There were the happy hours too, when with my loved one, almost daily, I set out to visit the old familiar places of our home town, where the beer was cool and the wine was clear, and the music seemed to sing of our love. It was springtime in Hamburg and springtime in our hearts, and the enchanted hours and days flew far too fast.

The present seemed all absorbing, but when Lotti and I walked and planned about the future, how quickly the thoughts flew to the Northern bush, where the two of us were going to make our home. How vividly there came to mind the picture of our camps, the boys, the friends, the northern lakes. That was the real thing, and it would be perfect, to live it with Lotti. The family didn't take kindly to the idea of my leaving again. Temptations were put in my way to change my mind. An Uncle without children, needed someone to take over his farm, etc.

"No, I am going down North again," I said again and again. It was hard to explain Canada, many things that I had learned to take for granted were incomprehensible to the folks at home. The vastness and emptiness of the country, the fact that a man could just go out and stake for himself enough ground to swallow several estates, those were things hard to grasp by people used to the European conditions.

But that wasn't all. I had enjoyed a wonderful vacation, I had been lucky to meet my future mate, the wanderer had been home again, but he wasn't the same fellow anymore, who had started out years ago. The years in Canada had wrought a great change in my whole make-up. First, there had been the struggle to get a firm foothold, to learn the language, the different customs and attitudes, and then, by and by grew the feeling of being at home, helped in all this by new-found friends. I had come to enjoy the easy-going ways, the freedom of movement and above all the freedom of speech, and all this we had enjoyed in the greatest measure in the North.

In some respects, there was a grain of truth too in what Pat used to say, "Hell's bells, man, you get just as much freedom as you can pay for." There was quite a tendency to measure a man's achievement in life by the wealth alone he managed to pile up, somehow, and to say "More power to him. Nice work, if you can get it, and get away with it …"

But on the whole, Canadians were not awed by just material wealth, and most of them felt they could do as well as the next one, given the right breaks. There was no meekness in their make-up, and pomposity and snobbishness unfailingly call forth a sarcastic kind of humor that will whittle down the stuffed shirts to their proper size. Newspapers, politicians, and business moguls may put on their show for him, John Canuck will listen to it all, tongue in cheek, but he will not be taken in by it, "Look who's talking!" he would say. "Wonder what he gets out of it?" And in this lay the hope that Canada shall in the long run belong to all Canadians, and not become a sinecure for a small group.

Looking at Canada this way and comparing it with what I had learned first-hand in the country of my birth, where one party had managed to usurp complete control, aided, and abetted by big industry, fearing socialism and communism, and by the army hoping to regain lost prestige the choice wasn't difficult. The few weeks I spent in Germany had been sufficient to feel the constant threat of the concentration camp held over the heads of the people, who had lost the last vestige of freedom, who couldn't even call their lives their own anymore. They all had to fall in line, and march, march, march. I had made my choice before, I now had it confirmed. I chose the country where I knew that freedom still had a home and a chance to survive. I will be leaving Germany and Lotti behind.

The day of parting came and when the last "Auf Wiedersehn", See You Again, was said, I stood once more on the deck of the *S. S. Deutschland*, heavy of heart with the sorrow of leaving loved ones behind. But when the liner slowly pulled away from the "Alte Liebe", Old Home, it was as if a heavy load fell off my shoulders, the invisible burden of living in a country, where freedom had been bound and gagged.

Fog was lying over the North Sea and soon the coastline was swallowed up. I made my way to the bar, where a few fellows were busy drowning their sorrows. Before long I knew the life stories of quite a few of them. That morning they all sounded rather sad.

The Chief Steward finally rescued me. On my trip east we had had many a pleasant chat in his cabin, and it was good to see at least one familiar face. He asked me with whom I would like to sit during the meals.

"How about this table? You will have the company of two young ladies."

"No, thanks!" I replied.

"This table, with one middle-aged lady?"

"Sorry to disappoint you, but I would rather sit with elderly people, to remain unattached, you know."

"So, it actually is serious??" his surprise showed.

"Yes, it is. And for good. We had a glorious time together. We hope to get married next year."

"Congratulations! I understand now, and I think I know the right place for you. An elderly professor and his wife, you will like them," the Chief Steward suggested.

They were a charming couple indeed, very devoted to each other. It was their first trip across the ocean, to see one of their sons living in the USA, and with astonishingly youthful zest they were looking forward to the New World. This gave us plenty to talk about. But whenever the conversation veered towards political questions, especially those involving Germany, the good professor adroitly steered us back onto neutral ground. That had been a typical attitude in the country I had just left, where formerly intimate friends avoided making political comments, the fear had been driven so deep into their minds. I understood that, although it had been a shock at first. But here on the boat, between two worlds??

Many a turn we took around the deck, the old professor and me. Often, he was wrestling with some mathematical problem, which proved far too high for me. Then came a night when the liner was making its way very slowly through dense fog. Iceberg warnings had been received, the foghorn boomed incessantly, and the ship's bell was ringing. The thick fog seemed like solid greyish white walls sliding past the railings. The professor had a novel idea of determining the boat's speed, details of which I have forgotten, but I haven't forgotten the other story he told me that night.

"You may have wondered," he said, "why I tried to avoid politics. Soon we shall part, you will go to your beloved North, we shall go to the Middle West to see our boy, and then we shall return to our homeland. If we were younger, I believe we would stay with our son, he has often asked us, but it is not good to transplant old trees. We are making this trip now, because, and this is what I wanted to tell you, soon it may not be possible. I fear that Germany is heading for a new conflict. The people will have told you that all the arming is done, solely to prevent an attack by Russia. That is the propaganda line. I say it is preparation for a war of conquest and revenge, and there are many who think like I do, but they are the older people. We are told today that we don't understand the new times, that we are superfluous, not wanted, yes, even that we are to blame for Germany's defeat in the last war.

"For almost forty years I have been instructing boys. I have tried to teach them not only what the curriculum called for, but also, I have tried to develop their ability to reason logically, to see all angles of controversial questions. I have also tried to instill into their minds the respect for the greater, eternal values, tolerance, honesty, and sincerity in their relations to their fellowmen. During my lifetime I have been through two bitter and unnecessary wars, I know what war means. And now… It doesn't seem very long since one of the new leaders, who think they can scorn everything we held in high esteem, sat on my knees, a nice little boy, my nephew. My name tells you whom I mean."

Sadly, the good Professor Hess shook his head, abruptly said, "Goodnight!" and went below, burdened with the knowledge that another Armageddon was in the making.

Thicker and thicker grew the fog. The engines were stopped, the liner was wallowing in the Atlantic swells, till finally the following afternoon the sun broke through and the westward journey was resumed.

New York again. Blizzard the last time. Now tropical heat was rising in waves, as I looked down on the big city from the top of the Empire State Building. Reunion with Dorothy and Uncle Henry with many a long, cool drink, till it was train time, and then the long journey back to Edmonton.

CHAPTER XIII
BACK TO THE BUSH

ALBERTA SUNSHINE AND summer heat greeted me when I got off the train. This didn't prepare me at all for what my friend the flying ace, Wop May, had to say when I showed up at the Canadian Airways' office, soon after my arrival, anxious to head North on the first plane.

"Not a chance of getting in there now. Break-up is late. Only a few days ago our last planes came out. You could have stayed another month on the other side."

"A fine time to find that out," I complained.

"Too bad, but tell me how was the trip? How are things over there? Are they really hell-bent for war?" and so a long talk was on.

Time on my hands and not much to do. In Edmonton, the days got long that way, and I was always looking forward to the evenings when friend George might show up, to invite me to his home or for a ride, and I had to thank him and his charming wife and their friends for many pleasant hours.

After several weeks of waiting for the resumption of flying operations, here we were already in the month of June. Pat wired me to wait for him in Edmonton. He had to come out on business. So, with more time on hand, it was a heaven-sent opportunity that brought me the fulfillment of a long-cherished desire: A trip to the Canadian Rockies!

George showed up one night with two of his friends, Allan, and Jimmy, whom I had met before. "These two fellows are going to Jasper National Park tomorrow. I told them you'd be glad to go along."

Just like that. The true Western spirit, no formalities, I liked those two keen young chaps and they apparently liked me. "Sure, come along. You have nothing better to do," invited Allan.

"Nothing better, I should say not! For years I have been wanting to see the Rockies," I answered.

"OK then. Tomorrow morning."

A lovely summer day was dawning when the three of us set out. Before long however, it grew hot and hotter. Around noon it was tropical. Several times the car stalled. Vaporlock. Thick, lazy, yellowish clouds of dust hung in the shimmering heat over the graveled highway. But all this couldn't impair our spirits. My two companions kept up their incessant bantering, reciting poetry, snatches of plays and trying to surpass each other in the use of the most fanciful language.

I tried to keep up with them, not always succeeding, but certainly acquiring some flowery additions to my knowledge of the English language. The sun was in the west, when entering the foothills, we saw the Rockies for the first time. Just a silvery outline on the horizon. Dipping into the valleys we lost sight of them, but when we climbed onto the crest of the next hill, there they stood pointing higher and higher into the skies. But the full magnificence, the grandeur beyond description, beyond mere words, was only revealed after we entered Jasper Park, when the setting sun was painting the peaks with flaming colors, and long shadows filled the valley. A herd of free horses thundering across the highway made us stop. We got out of the car, and silently let our eyes drink in the wild unspoiled beauty of the towering monuments to the Power of creation.

When we reached Jasper, night was falling. After supper and a few beers, we soon sank into our beds, tired but happy travelers. Sunrise next morning saw us walking through the still deserted streets of Jasper. We had till noon, when Allan expected to finish his business affairs, to engrave into our memories the glory of the Rockies, and again and again our eyes turned towards the mountains, glowing in all colours in the morning sun.

We wished we could have stayed longer, and we were a pretty quiet bunch as we traveled eastward again, into the heat and dust of the flat lands. Nearing Edmonton, I began to wonder if Pat had arrived in the meantime, what news from the North he would bring, and friend Jimmy had his mind on a story for his paper. "I hope there has been a nice juicy brawl in the north-end, or something like that. Easy to write up. I got to have a human-interest story before the deadline."

Suddenly we saw, far off to the south-east a mighty swirling black column racing across the countryside. A twister! "There is your story!" said Allan. The menacing apparition seemed to travel toward the highway, and it wasn't long before we had a good picture of the results of one of nature's caprices. Power and telephone poles broken like matches, repair crews already struggling with the tangled wires, a collapsed barn, another one with the roof lifted off. Jimmy was busy asking questions of the linemen and making notes.

"Won't take me long to write this one. Headline: Roving Reporter chasing cyclone, or something."

A few days later Pat and his family finally arrived at Cooking Lake where Wop and I went to meet them. At last, the time of waiting was coming to an end. It was good to see the friends again, to hear news from our camp, and to know that within a few days I would be heading north once more.

Shortly after Pat's arrival, the room at the King Edward was filled with friends, pilots, and prospectors, talking, smoking, and consuming large quantities of the drink of the day, orange and grapefruit juice and gin. After the first rush died down, the talk turned to political aspects, and somebody said, "Let's hear about your trip. What did you see over there? Are they going to war again?"

So, I told my story, what I had seen and heard. I told of all the people in uniform, the new barracks that had sprung up, masses of airplanes, new airports, the fears of the old people and the fanaticism of the young, and I made the then rather daring statement that it was high time for the other countries, and Canada too, to see to their defenses.

Some were thoughtful, but others said, "You are a victim of propaganda too, or you drank too much German beer."

There wasn't very much I could do about that, I only reported the impressions gained, the conclusions were my own. But I remember Wop saying to me, "To you it means war? When?"

"There is only one hope: a revolution in Germany, but the chances for that are slim. Revolutions usually come after lost wars, not before. If that doesn't happen, there will be war in three or four years."

Time now seemed to fly, and we began talking of the trip down north to get going with our prospecting. However, the best-laid plans…

One morning after a short trip in a taxi, without any previous warning, a blinding pain almost made me curl up on the sidewalk.

"What the hell is the matter with you?" Pat looked at me. "You are green in the face!"

"That's how I feel. I'd better go and see a doctor."

After seeing the doctor, the verdict was that my appendix had to come out. I swore and raved: "This was a fine time!" Wop, who happened to be present suggested I see Dr. Terwilleger to make sure. The verdict was the same: "It may pass, but I can't advise you to go north!"

"When can it be done, Doctor?" I asked.

"Tomorrow, if you want to. Go to the hospital tonight."

"Thank you, Doctor, that's fine."

But I didn't feel fine. I was a pretty sad specimen when I returned to the gang at the King Edward.

"What happened now?" asked Pat.

"Hospital tonight! Appendix out tomorrow," I answered.

"Well, for Christ's sake man, that's nothing to be blue about. I'll get you the best goddamn redheaded nurse in Edmonton!" Pat promised.

And what did I see when I awoke around noon the following day? Next to my bed there sat a strapping young lady in a nurse's uniform with the reddest head of hair I'd ever seen. Pat had been true to his word.

Soon after he showed up himself, grinning widely, "How goes the battle? How's the redhead?"

"Everything's fine. Just turn around!" Behind him stood the nurse with the red halo.

"Holy mackinaw. I didn't think the old Doc would take me literally! Well, if she doesn't get you up in a hurry, I don't know what will."

CHAPTER XIV
LOSING A FRIEND

TWO DAYS AFTER I said goodbye to the Royal Alexandra and the redhead, we were heading north in Mackenzie Air Services' old Fokker, piloted by Leigh Brintnell himself. While Leigh continued on his flight after refueling at Fort McMurray, carrying Maud, her little boy Pat and maid, Pat took Reg and me in our own plane.

Once more we followed the Athabasca down North. It was a lovely sunny day, fleecy white clouds sailing before the north wind. Upstairs it was somewhat bumpy, but that didn't bother us too much until we reached the south shore of Lake Athabasca, when it really got rough, and an ever-increasing headwind almost fought our plane to a standstill. For a longish time, it looked as if we were never going to get away from the mouth of the Williams River, in spite of our furiously hammering motor. Up and down, and sideways it went, but forward only slowly, oh, so slowly. With one hand I held onto the seat, with the other onto my still taped-up side, which didn't feel any too good. Even old Reg's nose looked somewhat longer than usual. At long last we saw with relief the Goldfields settlement below us. My legs were definitely wobbly when I crawled out of the cabin. Pat was somewhat weary, after that long fight against the headwind. He glanced at his watch, "Do you know how long it took us? Over three and a half hours! Bet you we haven't got much gas left either."

Under normal flying conditions it used to take us about two hours, this time we had almost doubled the flying time and gas consumption. Reg declared there was only one gallon left in our tanks.

While we were refueling, friend Bob suddenly appeared on the scene. Quite unusual for him, he had a worried expression on his face, and although his face lit up for a moment when we shook hands after the long separation, I had the distinct impression that something was radically wrong.

"What's the matter, Bob? What are you doing here?" I asked him.

"I am going outside," he replied indicating that he was leaving camp. "Lots of things are wrong!"

"Ah, come on, if it's something that happened in camp, I am sure we can straighten that out, let's go and see Pat."

"No, I am sorry, that won't help. My mind is made up. Maybe later I shall write to you, and give you, my reasons."

All my attempts to make him talk just brought the same reply, and even Pat didn't succeed to get it out of him. Just "you will find out for yourselves." This was deeply puzzling, and to me very sad news. I had come to like Bob tremendously and considered him one of my best chums. He was a splendid man in the bush, knowing it intimately.

In camp he was invariably cheerful, and always ready to reduce tough situations by making some wildly exaggerated statements about it. Many of his pet expressions had enriched our camp language. Last but not least, I shall never forget when he, all alone, set out on his 25-mile trek through the night to Goldfields to find out what had happened to Pat and Mack. Now he was going to leave us. Why?

I had an inkling of the answer when we landed at our camp. Something in the atmosphere was changed. The usual lightheartedness, the never-ending leg-pulling and horseplay were missing.

When I noticed that, I became more than ever determined to find out who had brought about Bob's leaving. It was good to see the old friends, Ronny, Art, Punk and camp resident Ned again.

However, a few others, who had joined our outfit shortly before drilling operations had started, quite clearly weren't very glad to see me again. Something began to dawn on me.

CHAPTER XV
PROSPECTING AGAIN

DRILLING HAD BEEN suspended. Now preparations for the sinking of a prospecting shaft to intersect Drill Hole No. 17, the one we had the strike about, were under way. Pat had written to me during my vacation that the core from that particular hole had brought some very high assays, as a result of which the big boss had ordered the shaft sinking.

The setters' shack had been brought from the west side before break-up and had been turned into a wing for the original log cabin, making it into quite a roomy habitation. Colorful curtains, mats and furniture gave it a cheerful, charming touch, heightened in the evenings by a blazing fire in the fireplace built from mineralized rock. It wasn't difficult to feel at home again in the North.

Off and on, however, the one question crept back into my mind, demanding an answer: Why did Bob leave? What has happened to the pleasant comradeship that seemed to be characteristic of life in our camp? One afternoon Pat and I were sitting near the lakeshore, discussing the new season's prospecting. It had been decided to send two parties, and I had taken it for granted that seniority would determine who would be given the chance to do the actual prospecting. It came therefore as a surprise when Pat told me that the first team would be made up by myself and Ronny, and the second team would be made up of Woods and Steeds, two fellows who had joined our outfit later than some other good men like Art, Bob, etc.

"Who is running this outfit now?" I asked Pat after hearing the news. Disconcerting as it was, it might be at the same time illuminating.

"What do you mean by that?"

"I just want to know. There are other men in this outfit, who have been longer with us, who have worked well. Why aren't they given a chance to go prospecting? Bob would have been in line too. Why did he leave? You ought to be able to give me an answer for that one?" I answered.

"I don't know. He wouldn't say."

"Well, as far as I can see, the other two fellows know the explanation. He was in their way. They wanted him to leave."

Pat's usual self-assurance seemed somewhat shaken when he came out with further disturbing news. The two fellows who wanted to be prospectors didn't want to adhere to our original agreement to share eventual profits or shares from the companies that might in the future be formed to exploit our finds. They wanted to work independently from the other prospectors, and not even give a certain percentage of the shares to Reg, the airplane mechanic, as we originally had agreed to do. After all he shared the risks with us, and the proper maintenance of the plane was of the greatest importance.

"I didn't like it myself." Pat said after he related all this.

"Then why in hell did you agree to it? You are the Boss." By this time, I was good and mad about the whole thing.

"I don't like the way things have been going here at all. I don't like the change that has come over this camp, I only hope, we can restore the old friendly co-operation that existed before these two birds spoiled things," I said.

We talked some more, but I didn't succeed in getting Pat to renounce the agreement he had made with the two other fellows.

"Maybe I made a mistake. But the thing is settled. Those two are ambitious, they will work like hell to find something." There was a short, somewhat bitter laugh, "Give them enough rope…."

"…to hang themselves? Hell, and damnation, they will tie us all up. Remember that we prided ourselves on the nickname they gave us in Goldfields? "Pat's Socialistic camp?" Because we all worked together, with a will for the good of all, without too much bossing going on."

"You are too sentimental about that. This is a cruel commercial world. What are you going to do?"

"Ronny and I will stick to the original arrangement. Reg will get his share. I only make one condition, we will not have the other fellows any nearer than twelve miles to our camps, wherever they may be. That fair enough?"

"That's OK with me. I'd like Ronny and you to go to Ace Lake first, and prospect north from there. Remember the big, quartz outcrops we saw from the air?"

"We are ready to go any time. The sooner the better. How about our agreement?"

"You mean should we write it down? Do you think that will be necessary between us two?"

"I just wanted to hear it. So far, we have not needed written contracts. I hope we can keep it that way."

"Jeeskrist, but you are difficult at times. Come on! Quit worrying. Let's have a drink."

"After all this talk, I am dry enough. Let me say one thing more though, I am holding on to what I think is good and right, I am going to fight for it. That should appeal to the Irish in you."

"Holy Moses. Sometimes I think you had an Irish grandfather. Well, here's to Ireland."

"Here's hoping we'll see it someday."

CHAPTER XVI
CRACKING ROCK AGAIN

CIRCLING OVER ACE Lake, not long after the discussion Pat and I had, we had no difficulty locating again the massive white outcrops shining in the sun, and a sandy peninsula jutting out from the north shore became our chosen campsite. Quickly we set up our tents, and the same afternoon saw us hard at work picking into the big outcrops which turned out to be composed of a granular or sugar-quartz with no mineralization to speak of. However, its presence showed that, as the prospectors say, a lot of action had taken place which, in itself is considered a good indication that mineral might be somewhere nearby.

In ever-widening circles we explored whatever ground could be covered on a day's trip from our camp northward.

Ace Lake differed from many other lakes we had visited. No fire had destroyed the bush, farther inland trees grew to respectable size in the heavy, sandy overburden which filled the valleys.

Several small streams, draining large muskegs farther north, meandered through the sandy flats, finally finding their way to the lake to discharge their load of silt. This explained the many fine beaches and sandy peninsulas, in some places almost reaching from shore to shore. The west end of the Lake was completely silted up, the water only a few inches deep, and the thick growth of reeds looked from far like green pastures. Jackfish paradise!

The streams gave us easy access to a large area towards the north, but these otherwise pleasant trips didn't yield anything interesting from the prospectors' angle. Rocky outcrops consisted mainly of solid granite with only sporadic signs of iron pyrites. The same held good for the west, where massive granite hills rose abruptly from the lake level. There we discovered the outlet of Ace Lake, a deep cut gorge through which the foaming waters tumbled to the level of another lake farther west.

Day after day Ronny and I set out in our canoe to explore another section, systematically working our way around the lake. From early morning till late at night we trudged through the bush, over the hills, cracking rock and taking the odd sample for closer examination at camp.

At noon we made our way back to our canoe, to boil our coffee or make our tea, consume a big hunk of our own special bannock made from a mixture of white and brown flour, corn and oatmeal, dry milk and baking powder loaded with lots of raisins. Every second day I would bake two big pans full in our tin stove. It was a satisfying concoction, which kept fresh and moist for several days. At night we would make our mulligan stew from bully beef or canned sausages, dried spuds, and vegetables, spiced liberally, and a big kettle of stewed dried fruits was always kept full, to help us keep regular.

It was the summer of the thunderstorms. The mornings dawned brilliant, clear, and fresh, and we would start out, counting on a fine day. Not long after, however, the heat in the bush would grow oppressive, making the sweat and mosquito dope run, pleasing only the vicious pests humming around our ears. Then the thunderheads pushed up from the horizon, dark, threatening, with Sulphur-colored edges. Gusts of wind ended the eerie calm. Hell broke loose. Blinding flashes, roar of thunder echoed and re-echoed by the mountains, and at last the rains rushed down like cloudbursts. Often it caught us in the bush, sliding and slipping on the wet moss, scampering for some shelter. Sometimes we made it back to our canoe in time. Under the overturned boat we would sit it out, till the mosquitoes re-appeared in reinforced formation. The storm was over.

Repeatedly we had seen, but always too far away for a good shot, lone caribou bucks. Our appetite for fresh meat had grown strong, after all the bully beef and canned sausages. One evening, we had returned to camp early, another storm was brewing, we heard close by a rustling in the bush and the muffled sound of hooves on rock. Ronny grabbed the rifle and stealthily followed the sound. A few minutes later a shot rang out. Ronny reappeared.

"Got him. A nice fat young buck. Forgot my knife. You coming?"

Together we went to where a clean shot had felled the animal. Death must have come instantly. It was lying in the grass as if it had just gone to sleep. His coat was smooth and shiny, of an unusual dark color, almost black along the back. We were nearly through skinning it when Pat appeared very opportunely to share the meat with us. At least we could give him that much, our prospecting hadn't brought any results worth mentioning yet, nothing even worth a few sticks of dynamite. There was no point in wasting precious resources. It was a good thing that we weren't footing the bill on our own or our prospecting days would have been over.

Ronny didn't think much of the possibilities around Ace Lakes I had gathered from his short remarks. Never very talkative, he had grown more silent as the days wore on. In the evenings, when I was busy getting our meals prepared and doing the baking, he just stood in front of our cook tent, hands in pockets, teetering back and forth from heel to toe, whistling softly *Stormy Weather* and staring across the lake. All my attempts at conversation failed and brought forth only a few monosyllables in reply. By and by, I began wondering: Is he going bushed? That would be a hell of a fix, and I hoped more fervently that we would find something interesting soon. Then there would be more activity, and he would snap out of his gloomy moods.

During my visit to the Old Country my brother had presented me with a powerful pair of field glasses. One evening, I dug them out of my packsack, and idly studied the scenery, finally focusing them on the opposite shoreline. The light must have been just right, slightly to the east on the southern shoreline I noticed what seemed to be a series of quartz veins cutting through the darkish rock just above lake level. I couldn't be sure, so handed the glasses to Ronny, "Can you see some whitish streaks over there, running through the formation?"

"Could be," said Ronny.

"Let's go over there tomorrow. The more I look at it, the more it seems we might find a major break."

"All right with me!" Ronny said with a glint of his usual enthusiasm.

We had gone far afield from our camp in the past weeks, covered many a wearisome mile through the thick bush. Maybe it was time to change our tactics. In any case, the next morning saw us crossing the lake. Soon after landing we realized that we had gotten into a different formation altogether and going inland it became clear to us that a wide depression opened up toward the south. The white streaks which I first espied with my glasses, proved to be quartz veins, however, they carried no mineral. The country rock looked to us like Diorites and Gneisses. We finally had gotten out of the granites, which was a welcome change. A short way inland we ran into a large, dried up muskeg, only here and there small islands of rock were sticking out, most of them very solid, whitish-grey masses, hard to break into. Farther and farther south we went, travelling slowly, a few hundred feet apart. We were convinced that the country looked promising, but where was the mineral? Would we miss it with all the overburden?

Around noon we decided to work our way back. Ronny going slightly farther east, I somewhat more to the west. The day wore on, hot and sultry. Mosquitoes!! Another storm was brewing. Salty sweat, mixed with fly dope stinging the eyes. *Christ'a'mighty, what a life! Slugging through the goddamned bush on a day like this*, a small but persistent inner voice rasping away on raw temper. *Why not give up! Tomorrow is another day. Beat it to the lake and make for camp!*

Over there, what was that, sticking out of the muskeg? A few weathered-looking brownish humps. Misery forgotten; I swung my pick feverishly. Rotten rock flies. I got into the more solid stuff underneath, cracked open a lump and there it was, the sheen of mineral.

"Hey, Ronny, come over here," I hollered and whistled, and then my chum came barging through the bush. "Look at this stuff! Full of mineral! Feel the weight!"

"What is it? Sure, heavy enough!"

"Lord only knows. But sure, as shooting, it's some kind of ore!" Then we both were hammering away, then went on to the next outcrop.

"It's all the same stuff," Ronny said positively.

A dense cystalline, blackish mass with glittering silvery mineral and dull pinkish spots shot through it. Out came the sample bags to be filled, we blazed a few trees going back to our canoe, our steps springy once more, the weariness forgotten. At last, we had found something that looks like real ore.

In high good spirits we paddled home, to return early the next morning to our find, to do more picking and looking around for surface indications to determine the extent of the showing. Several days thus passed in great activity, until one evening we heard Pat flying in. By the time we arrived at camp we could see the plane tied up, and Pat crouching near our tent, scrutinizing our samples.

"Hello there!"

"Where did you get this?" asked Pat.

"Wouldn't you like to know?!" said Ronny.

"Come on, where'd you find this? This is pay dirt, or I'll eat my shirt," Pat was excited again.

"That's what we thought. It comes from the south shore, there seems to be quite a bit of it," said Ronny.

"What is it though? Never saw stuff like this before!" from Pat.

"As far as we can make out there's nickel in it. The Prospectors Handbook seems to think so," I said.

"Hope you are right. Whatever it is, it looks damn good. You say there is a lot of it?" Pat asked.

"We picked up outcrops over quite an area. Lots of overburden though, and to the south it runs into a big swamp."

"Tell you what we will do. I have Art with a gang at Regent Lake, doing some blasting. The other two fellows found an outcrop with copper and galena. Not big, but looks OK. I'll bring Art and his gang over here to do some stripping. We better stake some claims, too," Pat directed.

"That's OK. Did you bring the map I asked you for? We want to go farther south. I'd like to have that map."

"Sorry, I forgot it. Next time for sure. By the way, I had word from the big Boss. Neely Lake Mining Company will probably be formed. Things are looking up! You can wire for your sweetheart soon," Pat finally had some positive news of his own.

"Boy! That's good news!"

"Well, so long then. I'll be on my horse."

On our way to the south shore, the morning following Pat's visit Ronny, who sat in the bow, suddenly called out, "Look who is coming!"

Not far from us, heading for the north shore a big caribou buck with a fine set of antlers was swimming. We swung the canoe around and paddling furiously tried to catch up with him. That fellow could travel! We were almost back to the north shore before we got really close. He was in no mood to play with us. When Ronny tried to grab him by the tail, the whole rear end of the canoe lifted out of the water. Our canoe got a terrific wallop, we almost landed in the drink, and by the time we had regained our balance and composure, friend caribou was scrambling onto the rocky shore, where he shook himself and hightailed for the shelter of the bush.

Two days later Pat returned with the first bunch of the blasting crew. It would take at least five or six trips to bring the whole gang and the equipment. Racing towards camp after the fifth load, the plane suddenly lurched to one side, one lower wing cut into the water. We stood on the shore paralyzed for a few anxious moments, then made for our canoe to paddle at top speed towards the plane, whose motor was just ticking over. Pat waved us off, carefully pointed the plane's nose toward camp again, and gently feeding the gas, managed to bring his stricken bird to the shore. He seemed quite unperturbed.

"A little close for comfort!" was his only comment.

A broken bolt on the undercarriage had caused the mishap. Fortunately, it had held till the plane had lost its speed. It was a close shave indeed. The undercarriage was badly strained out of alignment, and it was quite a job to force it into its proper place again. However, in spite of the lack of proper tools, we somehow managed to slip in a new bolt, using some long poles and pieces of drill steel as levers. First gently, then with full throttle, Pat went up and down the lake, then came in for another inspection. Finding everything all right, he decided to take off for our base camp.

"I'll get Canadian Airways to bring the last load. Keep your nose dry!" And off he went, as if nothing had happened.

The last load for the blasting and stripping crew was flown in by a B&M plane, and with it came news that Pat had returned to Goldfields all right, but that his plane would be out of commission for several days.

After staking claims to safeguard our find, we went to work with pick and shovel, hammer, drill and dynamite, and the results were encouraging. The best-looking ore we found close to the muskeg, and here many trenches soon filled with water, making it impossible to gauge exactly the full width of the veins or ore shoots. However, this was only preliminary work, and it was equally important to find out the extent of the area over which the mineral occurrences could be traced. We were highly satisfied to see that it was not an isolated outcrop but stretched over several claims.

About a week of hot and hard work had passed when Pat came breezing in. It surprised me somewhat that the good results our work had shown didn't make the expected impression.

He had been excited enough when he saw the first ore samples. Now he seemed preoccupied and in a hurry.

"What's the matter?" I asked him. "Somebody after you?"

"Well," he replied, "I picked up the extension of this show about four or five miles straight south from here. It looks even bigger. I put Woods and Steeds in there to prospect it."

I was dumbfounded. "You did what? Have you forgotten our agreement? Nobody within twelve miles, that was my condition, and you accepted it. Holy mackinaw, but this is getting good!"

"Now, don't get mad. The fellows were through at Regent Lake, and I didn't want to leave the other place alone."

"We, Ronny and I, could have gone there. We must have been close to it anyway. Of course, we never got the map either. I can assure you that I am good and mad. I expect agreements to be kept, by myself and by others."

"I am sorry, but it's too late now," said Pat.

"That's easy to say, but it would have been a damn sight easier to avoid this. Don't forget I am still looking for an explanation as to why Bob left us. Furthermore, I shall have to explain to Ronny now that our agreement is no damn good, after I first talked him into sticking to our original arrangement. You will look fine, brother," I fumed.

So, we had it out, hot and heavy.

"I suppose I shall never hear the last of this," said Pat finally, and that was the only thing we agreed on that day.

Ronny took the news as quietly as he accepted everything else, but in a way, he said enough, "What did you expect?"

Soon after this incident, the bull gang moved south, and Ronny and I built our next camp on the west shore of Nymark Lake, still farther south. We had picked the spot from the air, a flat, rocky peninsula, jutting deep into the fairly large lake. Had we known that our canoe wouldn't be brought in, we would never have chosen that campsite, because it later meant an awful lot of extra walking. B&M moved us and the most necessary things in two trips, the third one never materialized. We waited out the first day. When no plane appeared we started prospecting, and then we realized that without our canoe we were in a hell of a place. First, we had to get off the peninsula, and then we faced a steep range covered with jungle-like undergrowth. Fighting through it and going uphill had us panting, and swarms of black flies crawling into eyes, ears, nose and under our shirts didn't improve the situation.

This had to be repeated every day, if we wanted to get inland and from day to day, we kept on hoping that a plane with our canoe might show up. The days were still hot, but the nights were beginning to be chilly, and every morning I left written messages in camp asking for my heavy sleeping bag. Nothing happened for a long while and by the time Pat finally showed up in the company of Doc, flown by B&M, we were pretty short of grub and all they had to offer was a shot against typhoid fever. Doc thought he had a few suspicious cases in Goldfields.

If nothing else, at least we could talk things over, and decide on our future move. Outside of an interesting outcrop in the valley west of our camp, the stay at Nymark Lake hadn't netted very much. We had blasted and sampled this particular outcrop thoroughly, that was all we could do for the time being. If the samples should kick with something good, we could always return. Now we were ready to move again, and only a very strong south wind prevented us from doing it the same day. The wind grew to a storm that lasted three days, and when it finally abated, a B&M plane, with Reg on board to give us a hand, blew in and took us and our paraphernalia to Mackintosh Lake.

Reg didn't bring any grub, only a batch of bad news. We knew, of course, that he specialized in looking at the gloomy side of life, but that Pat needed a new motor for his plane, dashed our hopes that we finally would get a little better service as far as provisions, mail and moving were concerned. All summer long we had battled bad weather, often our provisions had run low, and our work had often been made unnecessarily difficult by the lack of our canoe. A lot of valuable time had been lost, and now it would soon be September of 1936. The summer was almost over.

Newspapers hadn't come our way for a long while, and when we asked Reg what the rest of the world had been up to in the meantime, he told us that a bloody civil war had broken out in Spain. That was a long way from the Canadian North, but I never forgot Reg's report that forenoon at Mackintosh Lake. Somehow it was fitting, black low hanging clouds came rolling up as we started to put up our tents, right after the plane left. Hardly had we finished that job when the clouds started to unload. Rain and more rain, that was almost all to be said about our stay in that location. We tried to make the best of it. Whenever it stopped pouring for a while, we set out, only to return soon after soaked to the skin by new downpours.

Many wet and dreary days passed. At last the weather cleared and Pat showed up, with a little fresh grub (we were down to our last reserves), and a bright idea, "Saw a likely looking spot on my way over, not far from here. What do you say I take you in there? You will have to hike it back though, it's only a pothole of a lake I'll never get out with a load."

"Anything to break the monotony," Ronny replied.

"OK. Let's go!"

Pothole was the right name for that lake, even getting down onto it wasn't so easy. Once ashore, we liked the looks of the country, though, it was wild and broken up. There was plenty of mineral in places, unfortunately only iron. For a few hours we prospected together, then Pat managed a hair-raising take-off, and Ronny and I went camp-ward. Just before darkness enveloped the country, we staggered into camp. By candlelight we retraced our hike on the map. A lot of walking for nothing, we had gotten onto the Old Man River power reserve!

A week of rain followed, so we had plenty of time to rest up after that long, futile hike. Our plan to move came to naught as no plane showed up. The weather turned worse, rain turned into hail, hail into light snow, and when once more we were down to our last can of bully-beef, we set out for Moore Lake, where a new camp was under construction.

Prospecting on our way we hit the camp just when the cook was calling the gang for supper. For the first time I encountered Alex, camp-cook, philosopher, and as Art, who was also present, put it, "some sort of a half-assed doctor!"

Alex certainly was a wonderful cook and baker, and to Ronny and me, it seemed like heaven to sit down at a table loaded with food, the likes of which we hadn't seen all summer.

We enjoyed the good things which Alex put before us. Our looks must have told him how hungry we were. We needed no coaxing to work our way through all he had to offer, and our appreciation of his culinary masterpieces seemed to make him very happy. When our first ravenous hunger was stilled, however, Ronny and I began to wonder why we had been on such plain and often short rations all summer. Something in the organization of our outfit had gone haywire. As prospectors, alone by ourselves in the bush, we should have had, at least, all that the fellows in camp took for granted, and we hadn't had a fraction of it.

Actually, it had been the lack of food which finally had forced us to walk to Moore Lake. There were other reasons, besides the grub question, which made us wonder what had become of the old comradely attitude which formerly had flourished in our camps. Many new faces were around, we could understand that, but soon we found out that there were several cliques, between which the relations seemed somewhat peculiar.

Alex gave us a hint when he said, "A few of the fellows in this here camp have the gold fever," and later that evening Art put it bluntly, "Woods and Steeds think they have the world by the tail on a downhill pull. They talk about nothing else but all the shares they will get out of this. Makes a guy wonder. They didn't find it."

I had a pretty clear picture by then, and I wouldn't have needed the further illustrations offered by Woods and Steeds. Both approached me separately and brought the talk around to the question of shares if a company were to be formed. Both wanted to know, "Did we have a written contract with Pat concerning the distribution of shares? Wouldn't it be much better if we had one?"

They were definitely worried, and as far as I was concerned, they could stay that way. So, I told them frankly, "You guys made your own deal. We made ours, and that's the way it stays. If you are worried about getting shares, I can understand that, because this is not your find. Pat found it, because he looked for an extension of our find, and you fellows didn't want to share with us. That's all. If you now want a written contract, try and get it."

Our talks didn't last long. Besides, Ronny and I didn't have much time for long discussions. We had learned that a Canadian Airways plane could be expected the following day. We therefore got busy collecting a two-weeks' supply of grub to take with us to Pat's pothole, to the west of which we wanted to do some more prospecting before the season closed.

CHAPTER XVII
SEASON'S END

PILOT ARCHIE WITH his Junkers showed up the following morning. It took some persuading to make him take us to Mackintosh Lake to collect our stuff, and then drop into the pothole we pointed out on the map: "Why in the name of all that's cockeyed do you guys have to pick the smallest lake? I won't promise a thing. Even if I get you in, I'll never get you out of there again."

"That's OK. Just try to get us there. We'll walk out if necessary."

"All right. Let's go. Too bad I have to make a living flying crazy prospectors around!"

Several times Archie circled the little lake to which we wanted to go. From above it seemed like asking a lot to set the plane down on that small body of water, and it wouldn't have surprised us if Archie had said, "Sorry. No go."

Instead, he finally swung off to the south, and coming back very low against a strong north wind, slid down towards the water. For breathless moments it seemed as if the quivering wing tips would scrape the hillsides, but the master at the controls set his big machine down for a perfect landing and taxied to the nearby shore. Hurriedly we unloaded our outfit.

"Don't count on anybody taking you out of this puddle," Archie declared. "There isn't a chance for a take off with a load, unless there is a storm blowing straight from the south. At this time of year, it blows from the north."

With that he went to the north end of the lake and gunning his motor from the start, came roaring down-wind toward us. With a few hundred feet of water to spare, he yanked his plane into the air, the motor howling at full revs. Lurching from side to side the plane slowly gained altitude and grew small as it sped toward the south. For a few forlorn moments we looked after the disappearing airplane, filled with a feeling we had by now often experienced; somewhat lonely and left behind. We were left standing on the shores of some lonesome northern lake, amidst the rolled-up tents, sleeping bags, boxes of foodstuffs and whatever else a prospector drags around when he moves. We felt cast off and not yet belonging. Once the camp was built and while it was only a tent for sleep, to cook meals, or seek shelter on a rainy day, it was a home, a place to return to, and as the surrounding territory becomes familiar, the loneliness was forgotten.

This time we didn't go to much trouble putting our camp in shape. Two weeks at the most we could stay. By then winter would be upon us, we felt. The weather was unfriendly, cold, and wind-driven rains often forced us to break off our prospecting hikes to seek shelter and warmth in our tent. We had clung to the hope of making another find before snow would fall and called the pothole Hope Lake. Again, and again, we set out to make our dream come true. Scattered mineral we found, but nothing that looked really promising.

Thus passed two weeks, and we began listening for an approaching plane, thinking that Pat would pick us up, even if we had to leave our equipment behind. The nights were getting cold, and once more the supplies were running low. Day after day we heard planes to the north of us, several times we even saw them far off, but nobody came our way. After another week we were down to a few cans of bully-beef, some dried peas, and a little flour. Still, we kept hoping. The first snow flurries began to drift down. Then we had only dried peas left, and a little flour and we talked about walking back to the new camp. We knew that would be a hard trip, with the countryside soaked by the interminable rains. So, we tried to attract a passing plane by starting a big fire. But nobody saw it although we kept it going for days.

We were getting hungry, and in desperation Ronny set out with his rifle to shoot something we could eat. Outside of a few squirrels, we hadn't seen a living thing in the bush, and that's all Ronny brought back, a couple of squirrels. There isn't really much left of a squirrel after a 303 slug has torn through it. They didn't make much of a meal. We couldn't understand why nobody came for us. Clear enough instructions had been left at the new camp, and old Alex knew that we only took enough food to last two weeks. After that meal of stewed squirrel and peas, we decided to walk out the following morning.

We awoke to a changed white world. Overnight heavy snow had fallen and was still falling in big wet flakes. In spite of that, after a hasty cup of tea, we rolled up our eiderdowns, fastened them to our pack boards and left renamed "Hopeless Lake". The going was tougher than we thought. We slid on the wet rocks, stumbled over fallen trees, and the clumsy bedrolls grew awkward and heavy on our backs. It took us two hours to cover two and a half miles. We had twenty miles, at least, to go. This would never do; it would be better to return to Hopeless Lake. Maybe the snow would disappear again, maybe. So back to our tent which we had left standing. Tired, hungry, and mad, we unrolled our bedrolls, and crept in. At least we could keep warm that way, perhaps sleep a little and forget our troubles.

It was still snowing when we awoke. It snowed all day. Again, we started our big fire on a point near our tent, just in case.

A warm breeze was blowing the morning after. The snow melted rapidly. By night there was hardly any left. This was our chance. "Tomorrow we'll try again, and we are going to travel light. We'll just take our axes."

From our last flour and baking powder we made some bannock for the trip. Darkness came early that October afternoon, and as we didn't have any candles left to read once more our badly worn magazines, we crawled into our bedrolls, hoping for a good day to come.

As soon as we could see well enough, but long before sunrise, we left camp, heading in an easterly direction toward the Old Man River. Up to the river valley the territory was familiar, but it was difficult travelling as we had to cross several rocky ranges running north and south. In spite of slipping and sliding, often on all fours, we made considerably better time compared to our first attempt. The morning sun was glinting on the water as we reached the river.

We knew that the water would be very cold by now, after all, freeze-up wasn't far off. Wading through the river in some wide but shallow spot would be a chilling experience. This had given us the idea that it might be possible to drop a big tree across a narrow gorge just above the falls. We had visited the spot in mind once before, but at that time we weren't concerned with the necessity of building a bridge and hadn't weighed the possibilities. What had at that time attracted all our attention were those deep, smooth-walled holes which a much mightier river of the past, using hard boulders as tools, patiently, through innumerable years, had ground into the solid rock bottom. Way above the present riverbed, the potholes were now dried up, and the hard, polished, round rocks, whose grinding had caused these phenomena, were lying idly at the bottom.

Now the river forced its way through a cut about twenty feet wide, its water flowing at tremendous speed, then tumbling into a foaming whirlpool below. Right close to the narrows we espied what we had been hoping for: two big pines. This would be a cinch. Our axes bit into the trunk, and over went one tree in exactly the wanted direction, the tip reaching well onto the opposite shore. Our satisfaction lasted only the briefest moment. The branches were too long. The rushing waters got hold of them and down the gorge and into the whirlpool went our bridge.

"Let's try the other one!"

Again, the tree fell exactly as planned, and again the water began to pull it around. Ronny hooked his axe behind one of the lower branches, trying to hold it, only to see the tree swing around, tearing the axe out of his hands. Axe and tree went into the gorge.

We decided it would be wiser to go north following the river, hoping to find a shallow stretch to wade through. After an hour's march we found a good spot, the river was wide and shallow, its bottom strewn with big boulders.

"Just made to order. Let's go!"

Step by step, carefully balancing ourselves, we made our way through the ice-cold water, which reached well above our knees. Almost across, Ronny, who was then in the lead, slipped off a moss-covered rock. I was close behind. Instinctively reaching out to help him regain his balance, I also lost my footing and the two of us landed up to our necks in the drink.

There wasn't a dry stitch on us as we scrambled ashore. Shivering, we pulled off our boots to pour out the water and wring out our socks, and then we hiked on, almost on the double, to get warm. The sun came out. That helped. One more range we crossed to find ourselves in a wide level draw leading straight north.

The worst was over. Travelling at a good clip, we were quite confident that we would reach Moore Lake before the day was gone. Wind, sun, and the heat of our bodies had evaporated Old Man River water from our clothes, and we were comfortable again. By and by however, we grew awfully hungry. And even the well-soaked primitive bannock didn't taste too bad, but somewhat reminded me of old Dick's remark, "Baking powder! That's pizen. Makes stones in your stomach."

Tempting visions of Alex's well-set table arose in our imagination to hasten our steps. Looking at our map, we knew that we had covered quite a distance. Since early morning we had only rarely stopped for a few minutes' rest, Ronny picking himself another blade of grass to chew on, while I lit my pipe. On we went. We had hopes of gaining the south shore of Moore Lake while there was still daylight. If we succeeded in that, we might be able to signal to the camp at the north shore to come and get us by canoe.

About a mile from Moore Lake, we were suddenly hailed by Terry Mahoney, who would come to be a good friend, and a few other fellows at their camp.

"Hello there! What brings you fellows here?" asked Terry.

"Just plain starvation. We walked all day. Ran out of grub. What are you doing here?" answered Ronny.

"Staking claims."

"On what? You make a find?" I asked.

"Not yet. Staking is just on spec. You fellows made quite a stir with your find. The country is staked solid around your claims."

"Well, that's nice. See you some other time, Terry," Ronny said as he thought it was a good sign that our success had caused more staking.

"We'd better keep going. So long!" I added.

Soon we saw our goal before us, but there was no answer to our calls and whistles to the camp across the lake. We had to add another few miles around the lake to our long journey, and two weary travelers finally staggered into camp as the sun was dipping behind the hills to the west.

Straight to the cook-shack we went. Old Alex, busy at the stove, turned around as we entered: "Oh, it's you two. Did you walk out again? I thought you might show up!"

"It's us all right, Alex, only a few pounds lighter! We are hungry, got anything to eat!"

"Sure, sure! Just sit down. I'll fix you up. How about a piece of pie and some coffee to start with?"

Good old understanding Alex. Before we knew it, there was blueberry pie and steaming coffee.

"Take it easy now, boys. Take your time," counselled our good friend. We had come to the right place. "How far did you have to walk?"

"Must be thirty miles on foot, it is over twenty miles as the crow flies," I answered for both of us.

"That's quite a hike. How long did it take you?"

"Oh, about nine hours, I guess," groaned Ronny.

"Don't eat too fast now. Take it easy. There is more coming."

"All right, Alex, and thanks for everything. Your food is about all we have been able to think of for the last two weeks. But tell us, what happened? Why weren't we picked up? Something is damned wrong somewhere!"

"Well, for one thing Pat has had an accident. Hurt his hand. He's all right now and back at Neely. Archie damaged his machine at Goldfields, he hasn't been in here again. I have mentioned you fellows repeatedly, but nobody paid much attention to that. You know, some fellows around here are too busy figuring out how to spend the money they think they will make out of the shares, if and when they get them!"

"That's a long shot, isn't it Alex?"

"True enough. In my time I have seen many of those beautiful dreams come to naught."

"Is that all they talk about in this camp? Even if it is, that's still no reason to let us starve in the bush," I knew that Alex was not at fault.

"True enough, but I suppose it's just another case of 'man's inhumanity to man'."

Those words we were later on to hear quite frequently from friend Alex. Unfortunately, they only too often fitted the case.

The following morning Art showed us the work that had been done by his crew. We were quite impressed by the amount of cross-trenching and stripping along the strike, and the abundance of mineralized rock showing in the trenches.

"How did the Ace Lake stuff look before you left, Art?"

"About the same as this. Maybe even better in places. Too much overburden though. Here it's easier."

"That's good news." At least that much we got out of the summer's work, and if the diamond drills should confirm our hopes, we would have reason to be satisfied. However, Ronny and I knew that given half a chance we could have covered much more ground which had never been prospected before, thereby increasing the chances of finding something really worthwhile.

During the latter half of the season, we were seriously handicapped by the fact that we didn't get our canoe back after we moved from Ace Lake. Added to this was the inexcusable failure to keep us supplied with the necessary food, while there was plenty at the base camp. These and other things I was determined to thresh out with Pat at the first opportunity.

CHAPTER XVIII

FREEZE-UP

IN THE AFTERNOON we were lucky enough to catch a plane for Neely Lake, where we found a rather quiet camp, with only Pat, his family, and the shaft-sinking crew present. Our arrival seemed to cause a surprise.

"How are you? How'd you get here?" asked Pat who still had his hand bandaged.

"We are all right now. But what in hell was the idea of starving us out a second time?"

"What did you say? Starved out? You mean they didn't get any food into your camp?"

"Exactly. We had the first square meal in two weeks at Alex's last night," I complained.

"How did you get there? Anybody pick you up?"

"No, we walked it."

"Holy Moses! I am going to look into that. You see," holding up his bandaged hand, "I couldn't fly. But I gave clear instructions to supply you fellows and get you out whenever you wanted."

"Well, you must have given your orders to the wrong guys. However, we survived, but I'd like to make damn sure this doesn't happen again."

"It won't. I can guarantee you that. Come on over to the house and see Maud and the boy, Little Pat," big Pat sounded proud.

"How are they?"

"Fine, just fine."

And so they were. Little Pat was growing into a strong lively boy. The North country seemed to agree with him.

After I gave my report about our last prospecting activities, we had a good long talk about future plans, politics, news, war rumors, etc., and how it all might work out and what it might mean to us. This set the theme for many an evening to come. Winter was once more upon us, and with freeze-up not far off, we couldn't do much. Besides, we figured we had earned a little rest.

It had been decided to make Moore Lake our main camp for the winter, as we would be carrying out diamond drilling operations there, and at Regent Lake. So, we busied ourselves getting things sorted out and ready for the move.

Old Fred and his crew were still drifting towards the drill hole, or at least so they thought, using hand steel, the compressor never got working properly. I went over to have a look at the drift one evening. Pat had told me that he had begun to wonder why they hadn't intersected the drill hole yet. Acting upon a hunch, before going down the shaft, I lined up the standpipe of the drill hole with the shaft opening. This I thought would give me an idea in which direction the drift would have to run. When I arrived at the bottom of the shaft and could see into the drift, my directions seemed to be all wrong. This couldn't be so, after all, the survey had been made with a transit. But the doubts came back: they hadn't found the drill hole. Back in camp Pat asked, "Well, what do you think of our hole in the ground?"

"It's quite a hole, and the rock sure looks interesting, but call me crazy, I can't help thinking that they are drifting in the wrong direction."

"You sure get some strange ideas!"

"OK You can laugh at me if I'm wrong, but I think another survey should be made," I shook my head.

Unfortunately, I happened to be not far out, as a new survey proved. Not that it mattered very much, because shaft-sinking operations were stopped soon after freeze-up.

In our talks during the long evenings, we often discussed the whys and wherefores of certain things that happened in connection with our finds and the initial steps undertaken to find out whether we had a mine or not. After we had struck values in our drill holes, we were ordered to send our samples to B.C. for assaying, and not much later the prospecting shaft was begun. This left us with only one answer; the assay results must have been very encouraging.

Our belief was strengthened by the fact that separate companies were now being formed for Neely, Ace, and Moore Lake. This was promising, but about the real foundation for these actions we were still in the dark, and although we were hoping for the best, it was equally possible that it was just a scheme to recover some of the money spent on exploration work so far. Whatever the reasons, we figured we could only gain by it. We knew the bush, and we had learned by now to find the mineral, but most of us were just babes in the woods when it came to St. James Street and Bay Street manipulations. By and by we would learn about that too.

Meanwhile, we enjoyed the kind of life we had chosen, filled with strong hopes that we might improve greatly on it. By learning from past mistakes and weaknesses in our organization, a more effective way of prospecting should result. But then, there came moments too, when it seemed daring to make long-range plans.

The news that came over our radio grew more alarming. "The March of Time" on Thursday nights began to sound more and more like the march of hobnailed boots. Deep into the wee hours of the morning we wrestled with world problems, trying in our own way to get at the roots of the eternal political upheavals that beset mankind. Why had all progress to be drowned in bloody conflicts? Why the senseless waste of the best and strongest youth in invidious wars fought with ever-deadlier weapons? Widely divergent views uttered freely led to long and heated discussions, and often we lost ourselves, till philosophical thoughts would lead us into the dim realm of the infinite.

There was general agreement about one question, however, which to us, who had come to the North and fallen in love with it, was of paramount importance. When would this tremendous country stretching for thousands of miles to the Arctic be explored, settled, and thus safeguarded against possible invaders from foreign shores, who even at this time might be working and scheming for world conquest.

We felt that here was the greatest challenge ever to young Canadians, to revive and keep strong the spirit of the pioneers, and that a government conscious of its first duty, the welfare and safety of the country, could and should lend a helping hand.

A few warning stories were heard off and on, even in Ottawa, but they were lost in the endless babble of politicians dealing with trivialities, anxious only about the number of votes they might rate in the next election. In the meantime, exploration and settling were progressing slowly, far too slowly and the field was left mainly to a few big companies who, so it definitely looked to us, were more concerned about getting hold of promising fields to cut out competition, than about actually developing them. To those big outfits even the successful, independent prospector often had to turn, because the means were lacking to do the necessary assessment work on his claims, or to undertake the expensive proving up process by diamond drilling or shaft-sinking.

On the other hand, the prospector, working under contract or as an employee of a mining company only too often found himself deprived of his justly deserved reward after making a strike, by the fact that the mining company will not push the development once the claims are staked and recorded, and kept in good standing by sufficient assessment work.

The prospector might get his shares in a newly formed company, but these might never be traded on the market; he couldn't get any money for them unless the company who issued them in the first place bought back his shares for a song, when the poor prospector desperately needed funds to carry on.

These are facts which definitely called for correction, together with measures to limit the reckless gambling with the treasures known, and yet to be found in the North. In spite of scientific advances, the prospector is still the fellow who had to break the rock and make sure the stuff is there, fighting heat and cold, mosquitoes and flies, and often loneliness. He deserved a better break, a fairer deal. That is all he asked for as incentive to deliver the goods. Let the true story of The North be told. To most people, speaking of The North brought to mind pictures of a bleak, forlorn, desolate country, breeding ground of mosquitos and black flies, a forbidding climate, howling winds, and blinding snowstorms.

There was another side to The North, a picture of innumerable smiling lakes, of verdant green bush of evergreens, poplars, and birches. There was the sparkling, clear air, breathtaking, magnificent sunrises and sunsets, fantastically beautiful cloud formations, night skies like dark velvet studded with a million brilliant stars and draped with the magic veils of the Aurora Borealis.

The winters were long and cold, but after the lakes were ice-bound and the snow had fallen, the growing hours of daylight were sunlit, and the air was calm and dry. Dark silhouettes of the bush stood against the horizon, throwing long shadows over the glittering powder snow, and over it all was cast a soft bluish light.

That, to me, was the North.

Towards the middle of October ice began to form along the shoreline. A few days before Pat made his last flight to Goldfields, he had been champing at the bit for quite a while, because his injured hand kept him from flying. Shortly before freeze-up, a different plane flew over our camp. It circled a few times. Hearing the roar of the motor, we went outside to see what was up.

"Must be a new pilot, doesn't know his way yet," remarked Pat as we walked toward the dock, "hope he doesn't try to come in too fast, and too close to this bay. A down draft may get him."

"Look out! Here he comes!" somebody shouted.

Pat could just squeeze out, "Holy cats!" Then we saw the plane hit the water. It bounced once, plunked down again, one side of the undercarriage buckled, one wing dipped into the water, pulling the plane around. Not a minute had passed when two canoes raced towards the plane. Ropes were thrown to our men in the canoes, and then in spite of the near tragedy, we who stood on the shore exploded with laughter. Furiously our men paddled, the plane didn't move, the two crews pulled in opposite directions. Great hollering and shouting from the shore brought harmony into their frantic efforts, and soon, all hands heaving mightily, the plane was pulled onto the dock. The excitement was over.

During the last week of October, the ice spread shore to shore, but it took almost two weeks before the first flight on skis could be undertaken. It was a long-drawn freeze-up. Off and on the weather turned quite mild again, even in the second half of November we still had some rain.

CHAPTER XIX
WINTER AGAIN

THE END OF the month of November brought the real cold, Lake Athabasca froze over, and we looked forward to our first mail from the outside, which must have accumulated at Fort McMurray for almost a month. Letters, newspapers, and magazines by the bagful came into camp with the first plane and helped to while away the long evenings which at this time of the year began at 3 p.m. Winter tightened its grip, lower and lower fell the temperatures.

I recall the day when we heard the news of King Edward's abdication. All that led up to it, or at least, all we knew about it had been a hot topic in camp, the majority, rugged and outspoken individualists calling the free North their home, staunchly siding with the King in his fight against convention. A silent bunch was gathered around the radio when the dramatic end came, "At long last…." That night it was 48° below.

Cariboo were travelling south again. A few fell to our guns, steaks, roast and not to forget meat pie à la Dan appeared on our menu. The barking of dogs heard from afar in the still clear air announced visitors one afternoon. Mushed along by a tall, black bearded fellow, a dog team pulling a toboggan smartly trotted into camp.

We had heard that a couple of trappers had their winter camp about thirty miles to the North, our next-door neighbors in that direction. Here was one of them. He had just run over to see if we could help them out with some cigarette paper!

"Sure thing. Come on in, take a load off your feet. You are just in time for supper."

Next morning, well supplied with Zigzag, a few other treats from our stores and a bunch of magazines, he happily set out again on the return half of his shopping trip.

There were flights, almost daily, to Moore Lake, Pat keeping an eye on the camp building activities. All seemed to go well, only as far as Pat's own cabin was concerned progress for some reason was slow. I had been expecting to go to Regent Lake where I was to be in charge, but one day, Pat came, "Maybe you could first go to Moore. My cabin is only half finished, see if you can't hurry it up a bit. I want to move before Christmas."

So, I went on a somewhat ticklish mission, to speed up a job that others had started, who might resent being pushed, and who presumably thought that they had worked hard and done their best. I must say I was impressed with the size of the building and the workmanship on the log walls. But that was all there was, walls and a roof, on the inside covered with thick frost. One tiny stove was fighting the numbing cold in vain. Fortunately, friend Art was around; we were used to talking freely to each other, and he went into action right away when I told him that Pat and his family wanted to move before Christmas and suggested we get some heat into the building to dry it out and make working easier.

As there were no more stoves in camp, we confided our problem to Ernie the Swede, our blacksmith, a master of the art. In no time at all he transformed some gasoline drums into most effective heaters. Then the picture changed rapidly. Floors were laid down, partitions put up, and I could begin to build bookshelves, cupboards and even the planned kitchen. Within a few days we were far enough advanced to receive the future occupants.

Planes heavily loaded with household goods and provisions shuttled between Neely Lake and Moore Lake and two days before Christmas Pat proudly led his little family into their new abode. In the cookhouse, meanwhile, Alex and his helpers laboured long and late, not only preparing the regular meals for many hungry customers, but also to get ready all sorts of extra treats for the coming holidays. The camp was well stocked with food supplies, including halves of frozen beef but now we were waiting for the turkeys to arrive, for the Christmas mail, and for some fire water.

One day to go till Christmas, and a heavy fog rolled over the camp. Ceiling zero, and visibility practically none, "No flying, no turkeys, too bad!" we all thought or said. "Have to eat them for New Year's, surely on a day like this nobody can fly."

We couldn't believe our ears when, in the afternoon, with only about an hour of dim daylight left, rum-rum-rum there came the sound of an airplane. It approached, grew louder, was lost for a moment, and came back again. We all ran outside to hear better. Still, it came, went away, there it was again! Louder than ever. I heard Pat who stood nearby saying, with concern in his voice, "That fellow is in trouble. To fly on a day like this."

Silence for a few moments, then the hum grew into a thunderous roar, into a battering crescendo of noise, an immense dark shape broke screaming through the fog, and just missing the roof of the new cabin, slid onto the lake.

"Jeeskrist! Holy Moses! Son of a gun!" is all our bunch could mutter. That fellow sure took our breath away. "Who is it? We can't see a thing. The fog has swallowed him. Here he comes."

Close to the shore a Mackenzie Air Service Fairchild taxied. The door opened, out came grinning Alex Dame, the pilot.

"I thought you fellows wanted your turkeys, so I brought them over. The mail plane isn't in yet. Maybe tomorrow."

"How the devil did you ever find this place in this here pea soup?" Pat asked.

"Had to do a little flying around before I found a hole to slip through. I had a pretty good idea though where I was," came the nonchalant reply. That Alex sure was a cool customer!

We took it for granted that he would stay with us overnight now that he was safely on the ground. Nothing doing, "Thanks a lot, fellows, but there is a little party on tonight over at Goldfields. I'll be back tomorrow to bring your mail if it has arrived." Off he went into the fog.

The following forenoon, with conditions not much improved, he plowed through the fog again with our mail. "Surely you are going to stay here now, Alex, and have Christmas dinner with us?" Well, you just couldn't keep that fellow down.

"Not just now," he replied. "I have a few more flights to make. I'll be back tonight." He kept his word. Enticing smells filled the gaily decorated cookhouse, as the whole gang assembled there to partake of Alex's sumptuous Christmas dinner. The cook had outdone himself. There wasn't a thing missing, and with the excellent food, a few drinks and many a toast the spirits soared high. Songs and carols floated to the rafters, out of the doors and over the lake, no doubt for the first time since uncounted millions of years ago the Precambrian Range was born.

Thus, we spent our second perfect Christmas in the Northern bush. "This is the life!" many times it was said. Most of us felt that it offered all we could wish for: interesting, albeit often hard work, in or close to the great outdoors, good food, and the companionship of real fellows, friendships forged and tested in common experience. There were exceptions to this, to be sure, some who had nothing else on their mind but material values, to whom this life in the bush was just a means to an end, not a wholesome, satisfying way of living, the wealth of which they couldn't see, because there was no dollar tag attached.

The success of our Christmas party in 1936 immediately led to plans for a New Year's celebration. Pat and Maud were going to combine the official housewarming with a New Year's Eve dinner for the original gang in the new cabin. A lot of things had to be done to get ready, and my lot was to build more chairs and, most important, a big cross-leg dining room table. By this time, I had become the proud owner of a complete set of first-class carpentry tools, and it was more pleasure than work to turn out solid and well finished articles to fill the cabin, adding to the comfort of owners and callers. The big table was ready when the roast was done.

Candlelight and blazing logs on the fireplace threw a soft flickering glow onto the festive setting, reflected in the shining, laughing eyes of our happy group. We had come a long way, in a short time. It had been just over a year and a half since we first pitched our two tents on the shores of Beaverlodge Lake. In perfect comfort we sat there in the sturdy log cabin, outside it was near 50° below. A deep dark night, just the stars lighting the firmament with their steely blue fire. Silence and then out of the night it came nearer and nearer, a chorus of male voices, untrained, and not in perfect harmony, but it sounded good, this lusty salute to life.

"The boys are coming!"

"The whole gang with Punk in the lead, pumping his accordion, others blowing tin flutes and whistles which came out of Pat's surprise packages, ordered in good time from the outside, the rest shouting and singing marched around the cabin in single file. The door was thrown open, and in they came, sounding their grand finale, "He is a jolly good fellow!" Glasses were filled and raised.

"Happy New Year! Happy New Year to you!"

Yes, a New Year had arrived and had been duly welcomed. What would it bring us? We had good hopes, and in my own heart there was one special wish for 1937, that it would bring for me the reunion with my dream girl. The first steps had been taken to get permission for my fiancée to enter Canada, and we were counting on having everything clear by Spring to start married life in the bush.

With hopes, plans, and wishes stirring in our minds and sometimes expressed, we spent the first day of the year in a leisurely fashion, enjoying the shelter and comfort of our well heated cabins. We forgot that outside it was 52° below, as from the radio softly flowed the melodies born in milder climes. Waltzes from Vienna, Tango, Jazz, Rhumbas, and music from Hawaii.

CHAPTER XX
DRILLING CAMP

DRILLING EQUIPMENT AND the diamond drill crew had arrived at Regent Lake, and once more I packed my bundle to take charge over there.

According to Pat the camp was ready; all buildings erected and fitted out. It was with no little surprise that I found a state of affairs in no way satisfactory.

Woods who up to that time had been in charge to supervise the construction of the camp, was full of explanations, none of them very convincing. Too bad Pat had been taken in. For myself, I decided right then and there to get to the bottom of things. This wasn't made easier by the fact that, with one exception, all the men were strangers to me, and I would have to win their confidence first.

Right from the start I was under fire from the foreman, Ezra, who swore with fervor and competence at one and all, and at himself for having been so damned stupid to have come to this Godforsaken country instead of staying to enjoy the mild breezes of British Columbia. Looking around to take stock of the situation, I couldn't blame him nor any of the other boys for being dissatisfied.

Those boys had worked hard and long to put up the buildings, the logs for which they had to haul for quite a distance from a draw to the northwest of the camp. All they had in the way of tools were three axes and one bucksaw. With these they had managed to put up five fair-sized cabins, one to serve as cookhouse, the others as bunkhouse, office, and settlers' shack.

There wasn't a claw-hammer, hand saw, brace, and bit, nor carpenters' square or level in the camp. On top of that, they had run out of nails. That explained why they hadn't been able to lay the floors and put up the bunks. There were no heaters for the bunkhouses.

The cook-stove was put up temporarily, and everybody had to sleep in the cookhouse. Things were lacking all over. It was a sorry mess I had stepped into.

The first thing to do was to determine what was urgently needed to put the camp in ship-shape order. The boys, from experience, were skeptical when I asked them for suggestions to complete my requisitions. "You won't get that," they would say. "We have asked for these things for weeks!"

This was an uphill struggle, but I was to get things straightened out, and this I made clear to the gang in a little speech, "We are going to make the finest camp in the north country out of this. I am going to get all the material that's needed within a few days, I promise you that much, and I'd like you to give me a chance to prove it to you."

By the next plane, my long lists went to Moore Lake, with strict instructions to hand these to no one but Pat. With them went an invitation to Pat to come and see for himself in what shape the camp was. He didn't show up right away, but the wanted material started to come in.

After about a week the worst was corrected. Floors were laid, windows and doors properly fitted, camp heaters and bunks installed. The whole gang was in better spirits, and it was high time, because drilling was to start, and now at least the fellows would have a decent and warm place to return to after their shifts on the drill or the pipeline. The finishing touches could be taken care of later on, when spare time could be devoted to my hobby: carpentry. Our main concern was now to get the drilling started. A few more loads of water-pipe, drill rods and gasoline were flown in by our old friend Santa Claus Alex Dame.

A narrow arm of a lake south from Regent Lake was closer to the drill set up than Regent Lake itself. Drilling supplies had therefore been landed in this inlet, which definitely was not an ideal landing or take-off spot.

When Alex was landing with a cross wind on his last load, the runway proved slightly too short and his heavily loaded plane skidded up onto the rocky shelf of the shore. Not far, but it was enough to damage the under-carriage somewhat, and to put a decided crimp into the tips of his propeller. Even that didn't stop Alex for long. Pieces of drill casing cut to right lengths were lashed with a lot of haywire against the weakened struts of his under-carriage, the propeller tips were straightened out between two sledges, and Alex took his plane once more into the northern skies, to head southward for more expert repairs.

We didn't see Alex Dame for quite a while after that, but by and by a story came back to us that he had been grounded. Not for flying his damaged plane out to Edmonton, so the story went, but because he had taken a passenger with him. Half-way to Fort Chipewyan they had lost the propeller. In spite of that Alex had managed a safe landing, only to find out that his motor was hung up by one lonely bolt. That's the story I heard. Knowing Alex, it wouldn't surprise me if it were true. Nothing ever seemed to perturb him.

I can still see him today, as he looked to me one day when we had run into thick billowing fog on our way to Moore Lake. Hunched over, almost as if he were asleep, he was sitting at the controls, just once in a while turning his eyes to right and left, looking for a hole in the pea soup.

When he found one, forward went the stick, hard rudder, and in a screaming sideslip we went down, to straighten out in the last few seconds for a smooth landing. Nothing to it, apparently. There was only the ghost of a grin flashing over his round good-natured face. He loved flying and he was good at it.

After New Year's, the temperatures had moderated somewhat. At night they would go down to 35-37° below, but the days were bright and sunny, and the air was calm. One fine day followed the other. The first hole was going down near the find. A long pipeline with two big heating coils built in pits, where fires were kept going continually, brought the water to the drill from the lake. The pump and another heating coil in an old gas drum were set up on the ice.

Two men were busy all the time cutting enough firewood, and as our woodcutters had to go farther and farther inland, a fellow who owned a dog team was hired to do all the hauling around the drill and camp.

Life now settling down to a routine, and it was very pleasant. We had a good camp, with comfortable quarters, kept clean and orderly by Ted, our bull-cook, and also, we were very lucky in having an excellent cook and baker in the person of Garfield, who kept his kitchen spotlessly clean, served excellent meals, and was untiring in inventing new variations in our fare.

Garfield's days were long, starting at 5:30 in the morning to get breakfast for the day shift ready. Shortly after that the graveyard shift would return to camp. After that, lunch for the camp crew, lunches for the four to midnight shift, supper, and often late snacks again. There was always plenty to eat in the cookhouse, and always new and different things served with a smile.

A lot of credit is due to Garfield for his important share in making our camp the finest in the North. I had had my dark moments when I first came to Regent Lake, but now I knew that here was assembled as fine a bunch of fellows as ever was brought together. Work, though it was hard, seemed play to them, there was no need for bossing, each one did his part, and without having to be asked. Our cook always had voluntary helpers with his multiple tasks.

In this friendly atmosphere even tough old man Campbell, our bit-setting expert, began to soften and mellow somewhat, and his swearing grew considerably less violent and less frequent. Twice a day he had to let off steam, though, it had become a sort of ritual with him.

The first outburst would be directed against the poor fellows coming off the graveyard shift. They would bring him the worn-out bits and tell him the footage drilled. This would get him going, and what he didn't tell them wasn't worth mentioning. "Jeeskrist and the cows come home, of all the goddam bloody nincompoops and sons of sea-cooks that it ever was my misfortune to be thrown together with in this God-forsaken country of barren granite hills, you take the cake. What in hell do you think this is, a damned maharaja's pastime, throwing diamonds away the way you so-and-sos go about it. Are you crazy? Or am I crazy ever to have come to this frozen-over bit of hell?" and so on and so on.

There was nothing wrong with the old man's mind, to be sure, but nobody took him too seriously. One morning his sermon was especially loud and long-lasting. So, I decided to find out what had happened. I was just going to open the door of the old man's shack, which I had fully expected to be blown off the hinges, when out came Bill Thomas, drillers' helper. He looked at me and shook his head. I never did find out what this day's rant was about.

Most of the day Ezra would be fairly peaceful, carefully setting new bits. At lunch-time Karl, the dog-teamster would often get it in the neck. Some mornings the old fellow would take a walk to the drill set-up. Following the trail across the lake, well packed by men and dogs, he would work himself into a stew over the solidly frozen droppings of Karl's dogs which bothered him, through the thin soles of his mukluks, and that would be just too bad for poor Karl, a rather quiet chap who didn't always have the answers.

All this was nothing, however, compared to the evening performance. Every night Ezra talked some fellows into playing a game of bridge with him. He would be so busy telling his partners all about their mistakes that he forgot his own cards. Inevitably he was beaten, and then he turned his wrath against himself, "Jeeskrist and the cows come home! Why don't you fellows take me down to the lake, drill a hole in my head and let the brains run out, if there are any, and fill it up with what the dogs dropped. Why don't you guys hit me over the head, and be done with it? If I never play bridge again, that will be too god-dammed early. Take me down to the lake, I tell you...!"

Most of that speech he would hold with only himself in audience. The boys left him to his ravings, to wander over to the cookhouse, or join the nightly gatherings at my cabin, where a comfortable couch and chairs had been built. Books and magazines we had aplenty, and we finally got a powerful battery radio too, which gave us wonderful worldwide reception. Often, we would just talk. By and by we learned the life history of everyone, and we got to know each other very well. As the last one, old Ezra would drop in, now and again in quite a friendly mood, and often it was he who told the best stories.

The drilling was progressing without much of a hitch, but the results were disappointing. Red, solid granite with only the occasional bit of iron pyrites. Where were the galena and the chalcopyrite of which the lead was to abound? The survey for the drill holes was made before the snow fell, when the rock and veins could be seen. Where were they? We tried deeper holes: no results. We changed the angle at which we were pushing the drills down: no results! Pat appeared one day, "What are you getting in your core?"

"Nothing, just solid granite!" I replied.

"That's impossible!" Pat blurted.

"Look at the samples. The survey is cockeyed, I'll bet you, or the lead was never properly established!"

Pat refused, "Can't be. It was prospected and surveyed before the snow."

I tried again: "Just the same, we are not cutting any veins. I'd like to put a vertical hole down right on the spot where the original find is supposed to have been made. We can locate that because they did some blasting there. Maybe that will give us an idea."

Pat was determined: "For now, better stick to the surveyed line-ups. Maybe later we can do as you say."

"OK But it's wasted money, I assure you. One more hole should be enough to prove my point."

CHAPTER XXI
STAKING TRIP

PAT HAD SOMETHING else on his mind. He had to go south on business. While he was away, he wanted me to stake some claims around Hopeless Lake. He still liked the look of that area, although our prospecting hadn't resulted in finding anything worthwhile.

Pat said, "Next year the country will be lousy with prospectors. If it is no good down there, we can let the claims lapse. At least, in the meantime, we'll have peace to look around."

"Well, the rock can't be any worse there than here. You will have to give me some men who have staked claims before and know their way around in the bush though."

"I will arrange everything before leaving. When do you want to go?"

"The sooner the better. Send a plane over to pick me up as soon as possible."

"All right, that's all settled then! Here is another thing. I want to tell you; the boss wants me to send a party to Yellowknife. There is a lot of talk about that country now. We will see a real rush this season down there. I have arranged for Art and Woods to go to Gordon Lake before break-up. We have talked about it before. You wanted to go, but now you'd better get yourself hitched first."

"Yea, I sure would have liked to go. I had no idea Ottawa would take such a long-lasting interest in my marriage plans or…"

"Don't tell me you wouldn't have started it!"

"Far from it. I would have gotten the Captain of the boat to marry us."

We had walked to the drill set-up; the last core was still red granite. When we got back to camp, Gar was just singing out, "Come and get it," and that, at least, still sounded good.

Pat had lunch with us and left soon after. On our way down to the lake, he remarked that he found the fellows rather quiet, and he didn't like it. Whatever the meaning of this was, it just didn't sit very well with me that day. The camp was in perfect order and the boys were happy, of that I was certain. If we were wasting time and money the blame surely lay not with us. That deplorable fact was a result of insufficient prospecting, and the substitution of hard work by tall stories. Pat was a victim and so were we, only we knew it. His remark, therefore, sounded to me somewhat like trying to minimize somebody else's shortcomings by finding fault with us.

So, I told him, "You don't know this gang very well. We don't see you often enough and they don't know you. After you come back, you'd better spend a week here, then you'll find out. Furthermore, I'd like to say that I don't consider it a proof for happy relations that in your camp everybody can call you a s.o.b. and you call everybody a bastard. This camp is OK and the boys here are OK."

"All right, all right! Don't get hot under the collar."

"I don't like unfounded statements, and I don't care to drill in barren granite as a result of unfounded statements."

There was more along that line, not pleasant, but the truth isn't always… I don't know whether it was sort of a parting shot, but when we got down to the plane, Pat had another bright idea. He was going to send me another man, who had just come from the city. A cousin or something of Woods. We could use another good man, getting enough firewood for the pipeline had become quite a problem, and I wanted a good man with the axe.

"Has this fellow been in the bush before?"

"I don't know a thing about him, Woods asked me if I would give him a chance."

"And now you peddle him off to me. Is that it? OK. The guy may be all right. He'd better be or won't last long here."

"Holy Moses, but you are crabby today."

"With plenty of reason. To use one of your own expressions, 'What do you expect me to do?' Kiss you? Better get me some good men for that staking job, and let them have the supplies ready, so we get there early."

"OK. Keep your nose dry!"

"So long! And good luck!"

Two days later a Mackenzie Norseman flew in and picked me up. In a few minutes we were at Moore Lake. Three fellows, new faces to me, were waiting there. But no supplies had been prepared, not even had they brought their bedrolls down to the plane. That didn't indicate much experience in the bush, but when I asked them, did they ever stake claims before, they assured me they knew all about it. I shouldn't have believed it.

In a great hurry we got supplies from Alex, the trader, who was his old smiling self when I greeted him. Too bad there was no time for a chat. A good talk with the wise old philosopher I would have appreciated very much that forenoon, but we had to stake claims, and the days were short. It was noon by the time we arrived at our destination, our erstwhile camp buried deep in the snow, but otherwise intact.

That afternoon we staked two lines, each four claim-lengths long. One of the fellows went with me. Our way led along the crest of a range to the west of the camp. The other two fellows just had to follow a valley which ran straight south from our tents. There was no mistake possible, and all went well. Shortly before darkness fell, we all arrived back at camp.

In an optimistic mood I expected to finish the job rather early the next day. It was by no means a complicated proposition, but to be quite sure all would go smoothly, I drew a sketch of the proposed claim lines. I explained the plan in detail to the other three before we started out at daybreak the following morning, "Colin and I shall go to the west, for two claim lengths, from the northernmost post on the line you other fellows ran yesterday. From that post you two go east for one claim-length. There you turn true south, better set your compass now, and putting in the No. 1 and 2 posts, you continue for four claim-lengths. Have a look at this map. You should end up right here on the banks of the Old Man River. OK? That's where Colin and I will meet you. All clear?"

"Yes, sure, one claim east and four south," agreed the other claiming team.

"Don't forget. Do not cross the river, stay on this side, and you can't go wrong," I clarified.

So, we parted. Colin and I travelled fast, the going wasn't too difficult. It was a fine, clear, and calm day. As we had a far longer way to travel and to blaze, we fully expected to meet our chums at the agreed-upon spot near the Old Man River. However, there wasn't a sign of them when we got there. We waited and waited, built a fire, made some tea, and waited some more. Nobody showed up.

The afternoon wore on. We followed the river going north, shouting, and whistling, and listening for any sound. No answer came from the white wilderness, no tracks could be found. We went south again. Nothing.

"Where in hell did those fellows get to?" Colin asked.

As the sun was nearing the horizon, we thought it wiser to return to camp, halfway hoping the other fellows might be there. The camp was deserted. We dug out Ronny's carbine rifle and fired a few shots to guide our wayward brethren home. Nothing happened, and somewhat worried we made our supper, all the time trying to find an explanation. Now it was pitch dark. The skies had clouded over. We knew there would be a moon later, but would that help? The next few hours grew long. Every few minutes we stepped out of our tent and hollered for all we were worth. The only answer: Silence. Around ten o'clock a light breeze sprang up, the clouds parted, and moonlight flooded the country.

"Come on, Colin, let's go and follow their tracks from where they started this morning. It will be better than sitting around here and stewing."

We got to where the others had put in their first post, and turned south from there, always following their tracks. I got out my compass: "They didn't go true south, Colin. Look at this, they went east-south-east!"

"Well, I'll be damned!" Colin was as surprised as I.

A few minutes later we got to the river, and there we saw it. The tracks were running straight across. Again, we shouted till we were hoarse and weak, and finally, finally, there came an answer, and we saw two dim shadows moving towards us.

"Hey, you guys, where have you been?" I asked.

"Lord only knows. We couldn't find the river!"

"Well, don't look any further, you've found it now. You are right on top of it."

Tired and weary, but happy and relieved that nothing worse had happened, we dragged ourselves back to camp. The following morning, we put in the last posts and in the afternoon the Norseman carried us back to the base camp. After fixing up the claim papers and sketches at Moore Lake, I had a good look into the supply situation. It didn't exactly surprise me to find many items, often requisitioned for our camp, but never received, and further it was a relief to find out that Western Steers still grew hind quarters. Garfield, our cook at Regent Lake had wondered about that. All we ever saw there were front quarters. Something was rotten somewhere, and my curiosity was fully aroused, as to who was behind these petty endeavors to withhold things from our camp when there evidently was plenty to go around.

I found a few answers myself, and old Alex the cook, filled in some details for me. He was a wise guy and not easily fooled by anybody, least of all by smooth-talking customers. The state of affairs that had developed at Moore Lake camp was aptly described by Alex's favorite saying, "the inhumanity of man to man," and as far as that goes, this seems never more pronounced when quickly to be gained riches are in question. It always came back to the same thing; some fellows thought they could grow prosperous in a hurry, and as their claims to wealth were questionable, they were afraid, and out of those fears were born the miserable efforts to make difficulties for others who had played the game fair and square.

But these weren't the only things, by any means, we talked about. As we sat there in Alex's private cubicle, a part of the cookhouse, our searching minds would try to penetrate the dense jungle of international politics, the news was getting more alarming all the time. To escape from the gloomy picture of mankind bent on suicide, we indulged in reminiscing, telling tales of our wanderings in faraway countries and sights never forgotten.

"Where," as Alex pensively said, "every aspect pleases, and only man is vile," and with that, we were at our starting point again.

Time to hit the hay, tomorrow would come early, especially for a camp cook.

CHAPTER XXII
JUST LIKE GOING HOME

A LOT OF new supplies, my bedroll and myself made up the load for the plane bound for Regent Lake. To me, it seemed like a real home-coming, to be together with my crew again, to see again the pleasant sight of our log cabins nestled between the trees, sunlight glistening on the bluish-white snow, and the smoke rising straight into the calm air.

Drilling was going on, still with the same negative results, and it seemed almost useless to keep on splitting the core, logging it, and making samples ready for the assayer. Every day, and many a night, I went across the lake to the drill to look at the core. To walk at night through the silent bush, with just the stars above and at times the moon lighting my way was, in spite of the cold, pure enjoyment. I would stuff a handful of chocolate bars from our commissary stores into my pocket, and often our cook would have a special treat for the boys on night shift, newly made pie or cake, some of which he would give me. Once on the other side of the lake I could faintly hear the Ford engine chugging away. At the end of the trail there shone the dim white light of a Coleman lantern through the canvas wall of the drill tent, and farther along the red glow of the fires along the pipeline. I listened to the drill and watched the return water pour into the sludge box. If it only would turn black! Then we would know that we were in the mineralized lead. But that didn't happen, there were no sulphides where our drill was churning.

Everything else was so right, the camp in fine shape, the boys in fine fettle, and the weather perfect. Since the New Year we had enjoyed clear blue skies, hardly any wind and temperatures between 30-35° below at night, a pleasant contrast to the winter before. A few days after my return from the staking trip, balmy breezes came blowing from the west. One morning, a Chinook had found us, telling us of the warm Pacific. Cabin roofs were dripping, and we had to pull rubbers over our mukluks for a few days, when winter regained the upper hand. Spring was still a long way off.

Fairly regularly planes of the two commercial airlines landed at our camp to bring supplies, fetch, and deliver our mail, and also bring us the latest news from the Goldfields settlement and other points north. This way word came to us that the first dance was to be held soon at Goldfields, and that quite a number of fellows from Moore Lake camp were planning to go. The pilot who brought us these tidings was going to fly the bunch from the other camp to the shindig. Did any of our fellows want to go? Did they?

Everybody seemed to have been waiting for just this to happen. Most of the boys had been in the bush a long time. Of course, we couldn't all leave the camp, some had to stay behind to keep the fires going, and this lot fell to old man Campbell who wasn't keen on going, "The Saint", who probably had moral objections against dancing, and myself. The Saint was that somewhat strange specimen wished upon us by Pat, whose Irish had been gotten by a hard luck story, I presumed. This chap had proven to be a square peg in a round hole. He was the only who didn't fit in, and never made an attempt to play along with the boys, but rather irritated everybody by keeping aloof. In the bunkhouse, he would sit on his upper bunk and read religious pamphlets, declining all invitations to a game of cards or cribbage, or even to join in our nightly gatherings around the radio. Thus, he became The Saint and finally was left strictly alone.

Repeatedly I had talked to him, trying to make him see that in such a small, isolated group it was necessary to play ball with the gang. All my talking did no good. If he had been a good worker, or at least had been trying hard to do his share and to get along, he would have been all right.

The day set for the dance was approaching, it had taken on the aspect of a real event. The boys hardly talked about anything else, but when the time for the flight to Goldfields drew near, joyous anticipation turned into apprehension. There was a sudden change in the weather. Thick moist fog rolled over the country. Would the plane come? Our would-be revelers grew worried, and the possibility of hiking it to Goldfields was discussed. A decision was quickly reached, when the night before the dance, a bunch of fellows from Moore Lake, with Art in the lead, drifted into our camp, "What brings you guys here?" I asked.

"We are going to Goldfields, are you coming?" Art responded.

"Better wait till tomorrow morning, maybe the weather will clear by then."

There was no holding them the next morning, with the fog still lying heavy over the land.

I suggested, "Better take some grub, and a couple of axes along. You have a long hike ahead of you. How about a map and a compass? Which way do you plan on going anyway?" I asked Art.

"I think we'll head toward Beaverlodge Lake. Hell, it's only twenty miles."

"Twenty miles by air! By the time you'll get there, you will have made thirty-five miles. If I can give you a piece of advice, based on my own experience, you'd better go toward the big lake along Old Man River. It's a little longer but you'll have level going all the way," I suggested.

"OK. We'll try that."

"Well, good luck and have a good time!" Off they went, full of beans, joking and kidding, leaving behind a very quiet camp. Late in the afternoon a strong breeze blew away the fog, and not long after a Norseman landed near our camp.

"Where are the fellows for Goldfields?" called Stan, the pilot, from the cockpit.

"They started out this morning on foot. Didn't think you'd make it! Maybe you could pick them up, I advised them to follow the Old Man River."

"I can only try. You fellows not going?"

"Somebody has to look after the camp."

"All right, here we go," and off went the big silver bird, flying into a glorious sunset.

Two days later a bunch of weary and disappointed travelers flew back into camp. "Did you have a good time? You look as if you had been through the mill? How was the dance? Who was there?" We plied them with questions.

"Well, we never got there!" When the story came out; some of the fellows had found the going much tougher than they had thought, and Art and Steve must have set a terrific pace in their anxiety to get there on time. Soon, feet not accustomed to snowshoes got sore. Some fellows got awfully thirsty, and arguments as to which way to go arose.

In the afternoon, a few had gone as far as they could. Finally, they decided to build a lean-to and a fire. Only Art and Steve went on. They got there all right at 3 a.m. and later in the morning sent the plane to pick up the others, those that now had returned.

"Where are Art, Steve and Karl now?"

"In Goldfields, there is still some celebrating going on; a plane from Port Albert came in with some booze. Last we saw of Karl he was heading toward Muskeg Myrtle's tent."

"You fellows better rest up today. Tomorrow we have to move the drill," I suggested.

Although the expedition had been a failure for most of the gang, there seemed to be a lot to talk about, and a lot was heard about a certain Redhead newly arrived in Goldfields, competition for Muskeg Myrtle. Apparently, she had great success in relieving the miners of their hard-earned dough and making them forget their loneliness.

I had been a witness to the first invasion of the new mining town by the oldest trade. Hearing the boys now brought back the picture. Pat and I had flown to Goldfields and were loading our plane with supplies from Alex, the trader, when a big commercial plane pulled to the dock. Passengers got off, among them a woman whose prime had long since passed.

"Whom did you bring?" somebody asked the pilot.

"Whom did I bring? This guy asks me! Brother, I brought civilization!" At that moment appeared at the door of the plane the mechanic wrestling with, lo and behold, the massive, shiny, yellow headpiece of an old-fashioned brass bed.

"What in hell is coming next?" said Pat.

Then came the mechanic's voice, "Hey, you guys, give me a hand with this here workbench!"

A few days later than the other fellows, Art, Steve, and Karl returned to camp. All looked considerably worse for wear, and at first, we all thought this was the result of their carousing. Soon we noticed, however, that with Steve it was more than a hangover, and the kidding stopped. Steve, the most powerful fellow in camp, who, according to his brother Ronny, no mean athlete himself, could just pick him up and put him down, had caught himself a terrific cold or maybe the flu. It was the first time anything like that had happened in our camps in two winters.

Soon after we got Steve into his bunk, a raging fever got hold of him. The poor fellow shook and shivered uncontrollably. A spare bedroll and all spare blankets were rounded up and piled onto him. Aspirin and 222's were the only medicine we had in our First Aid kit. Stiff doses were washed down with enormous quantities of orange juice and water. Ted, the bull-cook and I sat with the patient all night and the following day, keeping a roaring fire going to dry and change the constantly wringing wet blankets.

When finally, the fever broke, Steve, "the horse," was as weak as a baby, only a shadow of his former self, and his nurses were ready to drop in their tracks too. However, we had won the battle. Whatever had hit Steve, nobody else caught it. This was maybe partly due to our efforts to disinfect the whole bunkhouse, every movable object was put out into the fresh air, and buckets full of Lysol solution were sloshed over floors and walls. The main credit however, why the sickness didn't spread, must, I suppose, be given to the dry, clear, cold air of the northern winter.

It took Steve only a few days to get back his strength, but a little longer to make up for a considerable loss of weight. Gar, the cook, did his best to feed him up.

While Steve was still out of commission, one of our drillers had a seemingly slight accident. Carrying a length of pipe, he slipped on a rock and fell on his back. A few days later he complained of severe pains and had to go outside for medical attention. One of the helpers, who by then had learned the tricks of diamond drilling, filled the breach very satisfactorily, but it meant re-arranging our crews, and The Saint had to go on shift work to keep the fires under our pipeline coils going. He knew what it meant to keep the water flowing to the drill, but just to make sure, before he set out on his four to midnight shift, I gave him a pep-talk, "Make sure to keep the fires going, if you let her freeze up, you are through!"

Around 11 o'clock that same night, the drillers' helper came running, "The drill is stopped. No water. The pipeline is frozen solid."

This was a fine mess. Old man Campbell was furious and fumed, "Now, even you should have enough of that damned lazy bum. Get him out of here. Don't let me see that so and so again, or I'll beat the stuffing out of him. If he has any."

We hadn't heard him go that strong for quite a while, but we agreed this time he had good reason for his wrath. I was mad too, but partly at myself. I should have known better. It took a lot of hard work to get the pipes thawed. The next plane leaving our camp had a passenger; The Saint went home.

"What next?" some were asking. They didn't have long to wait. Karl came running the next morning as we were sitting down for breakfast, "The leader is dead! I lost my lead-dog!" The poor fellow was terribly upset, so much so that we got the idea he was putting on a good show, and instead of getting sympathy, he was in for a lot of kidding.

"Don't cry now! You never treated him right! You spoke indecently to him, we heard you! Did you say goodnight to him last night? There now, you didn't. You hurt his feelings, and he just curled up and died. Poor dog, but he's better off now, doesn't have to slave for you anymore."

Karl didn't know what to make of it all. That same night he was the first one to show up for our nightly meeting around the radio, he wanted to pour out his heart to me. First, he told me another long story about his leader, but soon he forgot that over another matter, "the Redhead". To get another lead dog he should really go to Goldfields, and why didn't I come along? The two of us could have a swell time, this Redhead now, she really was something, she was beautiful. He positively grew lyric about her charms, "The skin on her thighs is just like, like sammet. No, that's German. What do you call that in English?" He meant velvet.

"Well, I am sure you don't need any help from me there. You want to get next to the Redhead, I got that, but first you talked about a new dog! We got off that altogether."

"Yea, it's too bad there aren't two of them. But I tell you what we'll do. We go to Cannery Bay. Some of the women aren't bad either. Maybe I can get some fish there too, for my dogs. It's all that mush that killed the leader." At this point the entrance of some other boys fortunately put an end to the chapter on Karl's love life.

"What was Karl talking about when we came in?" one fellow asked.

"Oh, dogs again," I replied.

"I know what he can do. Why not train Windy as his new leader?"

This was as good as trying to tell Karl that the Redhead wasn't charming, "Windy, a sleigh dog. You are nuts," he laughed.

Windy had been left with us, by her former owner. She was strictly ornamental. Hers was a beauty like a thoroughbred Alsatian, and on this alone she got by. If she ever received any training, she never betrayed it. Orders didn't exist for her, at least she never obeyed any command. To get any work out of her would be futile, but to try it might be lots of fun. So far, whenever Karl harnessed his team, Windy had just been the interested spectator. She would prance around the poor working dogs for a while, then stand back in regal pose, looking down on them like a ravishing demi-mondaine might tauntingly look down on her hard-working brothers and sisters, just as if she were saying, "Look at me, look at my beauty. I don't have to work. I get along fine."

Windy won out the next morning too when an attempt was made to put her in harness. Karl had lots of assistants, but that didn't help. Quicker than they got her into the harness, she was out of it, or she got the whole team all tangled up, till it was just a snarling, barking free for all. In the end, she got away and kept out of sight, until she heard Karl's "Mush, mush!" Tail proudly flying, she came bouncing back, and went through her usual performance of prancing around the team, strutting her stuff, joyously barking, "See what I mean? I wasn't born to work."

Towards the end of March Pat finally showed up again after his vacation. All the time we had been drilling in red granite, with precious little mineral showing. I therefore had no encouraging news for him when he arrived.

"How are tricks, find the lead yet?"

"Everything is fine, but no mineral," I shared our disappointment.

"Holy Moses! We'll look fine when the big boss comes in. He is due in a few days. What you think is wrong?" Pat asked.

"As I told you before, no thorough prospecting, and a survey by guess, that's the only explanation for me. Too bad," I grumbled.

"Damn it all, sir. I saw the mineral before," Pat spluttered.

"Sure, if you had let me drill that vertical hole, we might have been wiser. As things are now, I am willing to bet that if there is a mineralized lead, it is running from the first show in a more southerly direction. After the snow is gone, we can find out."

"Well, that's that. How are you own affairs developing, when is the wedding?"

"The Lord only knows. I am deeply entangled in red tape. No final word from Ottawa yet. Last thing I heard, I had to put up a bond. That's done, but I don't know whether that was the last hurdle. We should have let the captain of the boat marry us," I sighed.

"Don't worry. If nothing else, I'll get my old man to pull some strings," Pat offered again.

"Go ahead. The sooner the better. I am getting worried. Time is running short."

"What do you mean? War?"

"Looks bad enough, doesn't it? But, even if we don't consider that, you know, over there even the girls have to go into labor camps, or work for farmers and Nazi big shots. I'd like to save her that."

"The dirty bastards! I am going to write my old man tonight."

"Thanks. It will take a load off my mind, to know that something is being done!"

"OK Keep your nose dry. I'll be on my horse. Be back as soon as the boss shows up," and Pat was off.

We didn't have long to wait. It was a fine, sunny morning when Pat brought the big Boss. Not a day for bad news, but I had no illusions that one look at our drill cores wouldn't be sufficient to terminate the drilling, perhaps close the camp for good. However, first there were friendly greetings and then an inspection of the camp.

I was proud to show it. We had made a place of it, orderly and attractive, to us it had become a real home. At last, we got to the core shack.

"How did the drilling go, Hermann?" asked the boss.

"We had no trouble with that, although Ezra says, 'the rock was harder than the hubs of hell.'"

"Where is the mineral?"

"We never got into it."

"But it was there?"

"Sorry to say, but not where we were drilling," I explained.

"Did you prospect here? Did you see the lead?" the boss wanted details.

"No, sir. I came here after the snow was on the ground. The survey for the drill holes was done before that."

"It's no use going on like this. A lot more prospecting will have to be done first. We better take the drill crew out for now," was the boss' instruction.

CHAPTER XXIII
END OF DRILLING

THAT WAS THAT. Now I would have to tell the boys, that was my first thought, and I didn't like the prospect.

Down to the lake and the plane we walked together. The big boss, a very silent Pat, and myself. Not saying much either.

Just before boarding the plane, our boss, one of the leading mining men in Canada who had seen many a bright prospect fizzle out, no doubt, very understandingly softened the blow somewhat. "Better luck next time, Hermann, we have to keep on trying. I want to say though, this is the finest camp I have seen in a long time."

At least we had succeeded in that.

When the plane had risen from the lake, I made my way to old man Campbell's shack, with a heavy heart and leaden feet. The old man, with one long grey lock falling over his spectacles was still busy setting bits, cranking his hand drill, then picking up a stone with a pair of pincers, trying it in the hole he had drilled. Not quite deep enough! A few more cranks of the drill. Again, he tries it. It's all right. The diamond was just slightly protruding. Now for the punch to drive the metal against the stone till it was firmly anchored. Only then the old fellow looked up.

"What's the news?"

"No good. We are to stop drilling and get the equipment ready to haul it out," I said sadly.

He looked at me, sadly and quietly, walked back to his bunk, sat down, and put his toil-worn, capable hands over his eyes to hide the tears that rolled down his cheeks. I could have bawled with him. I knew how he felt.

In the beginning we had battled a lot, but we had taken each other's measure, and I knew that under his rough manners was hidden a big and faithful heart. We had become friends and now we had to part.

"Let's not take this too hard," I finally managed to say. "Next time we drill together we'll be luckier."

"That's not it," he replied. "I liked it here. I was happy. Happier than I don't know since when. Never was together with a better bunch of fellows, for sure."

"I feel the same," I answered, "and I am very sorry too that this comes to an end."

"Yes, but you are young. At my age, every hour counts, and you know, out of the total the happy hours are the rare ones," old man Campbell shared sadly.

"That's true, and I shall try to remember that. Now, I'd better go and tell the boys," I replied.

We had all known, of course, that our camp could only be a sort of waystation in our wanderings to search for rock-bound treasures, but in spite of that, leaving it proved hard and sad.

First to go were the drillers, after the drill, and all that went with it, had been dismantled and prepared for shipping. Then followed our crew, some going to Moore Lake, others heading for Goldfields, Karl and his dogs making up the first load. While the plane was being loaded, I talked to Stan, the pilot. "Be lonely for you guys here. But look, he pointed up to the skies, where mares' tails were sailing high, "the equinoctial gales are coming. Spring will soon be here, and you'll go prospecting again, I suppose."

Finally, there was only the cook, Steve and me left in the quiet little settlement. We busied ourselves making a great spring cleaning of the whole camp just in case we too would move out in a hurry to take up prospecting again.

CHAPTER XXIV
BREAK-UP

PAT PUT IN an appearance and settled the question for us as to what our next tasks would be. But first he told us that he had received word from Yellowknife that our boys found free gold during break-up.

"You know," he said, "we sure took a beating here, but I still think the mineral is there. So, what I like you to do is to prospect the whole area again, and this time establish the lead without a doubt," he requested.

"If it's there, we shall find it. I have an idea what may have happened. Steve will then be a prospector from now on?" I asked.

"Yes, if you want him. I am taking Garfield with me."

"Steve suits me fine as team-mate. He is not afraid of work, and we shall have to crack a lot of rock. I'll bet you there hasn't been enough of that around here."

April 1937 had come, and with it, warmer weather. During the middle of the day, the snow melted a little, but the nights were still quite cold. Steve and I puttered around the camp, trying to make the days short, until we could go into the bush again. The old snow was almost gone, when suddenly, this was around the middle of the month, a violent storm strangely accompanied by lightning and thunder, blanketed the country with snow once more. Cold weather followed and lasted till the end of the month, when at last we were sure that spring was with us for good. Temperatures soared within two days from the freezing point to 72° above, and we discarded the heavy underwear.

Loneliness was with us in the almost deserted camp. To combat it we kept as busy as possible, but with the warm breezes of Spring blowing, intense restlessness got hold of us. Along the shoreline some rock was now becoming exposed. So, we grabbed our prospectors' picks, pulled on rubber boots, and began exploring, sloshing through almost foot-deep water on top of the ice with Windy prancing around us. Cracking rock frequently, we worked our way along the south shore going east. Prospecting was new to Steve, but not long after starting on our hike, it was he who called out, "Look at this! This must be mineral!"

"It sure is. Chalcopyrite and even Galena. This is the stuff we were looking for in the drill core. Where'd you get it?"

"Right here. Look. That whitish, glassy stuff looked like quartz to me. That's what you told me to look for."

"You've got a vein there. Let's try to follow it. Seems to run parallel to the shore."

That first day out we found enough mineral to give us a tentative explanation of what actually might have caused the failure of the diamond drills to reach the lead. "Look, Steve, this is pure speculation on my part, but if you take the direction in which this vein is running, and continue that line to the west, where do you get to?"

"South of that big hill. And we were drilling to the north of it," he answered shaking his head.

"Mind you this is just an idea, but it gives us a plan to work by, unless we find out otherwise. That big hill is probably new intrusive rock, very likely the red granite we got in our core. Where the first mineral was found is probably the fault, the end of the lead was possibly dragged around to the north. If we had drilled farther south, we might still be at it. We shall know more when the snow is gone."

Quite satisfied with our first day of prospecting, we turned for home. Halfway to camp, making our way around an island we suddenly found ourselves almost face to face with a bunch of caribou.

"Let's get our rifle. Fresh meat tonight!" The prospect was almost as exciting as our find at the fault.

Windy was still in camp with us. Seeing the caribou and giving chase was one with her. She paid no heed to our calling and whistling. In no time at all the herd was scattered from one end of the lake to the other, with the barking demon behind them. The frightened animals skated and slipped on the ice, desperately trying to reach the shores and better footing. By the time we got home and grabbed our rifles there wasn't a single caribou in sight anywhere. They were the last herd we saw that spring. So, it was bully-beef again. Windy came back, while Steve and I were sitting in the evening sun in front of the cookhouse, panting hard with her tongue hanging out, she put herself at our feet, looking from one to the other as if she were saying, "Boy, we sure had some fun today."

Steve growled, expressing my feelings too, "Damned useless bitch. Caribou steaks sure would have hit the spot tonight." Windy looked as if she didn't understand a word he said, and went off to sleep, twitching and whimpering once in a while, dreaming of the great hunt.

Warm, westerly breezes continued to blow. The ice broke from the shore and began to drift with the wind. As long as we could jump onto it, we used it as our highway for further explorations along the shores. Full of ruts and busy drain holes it was rapidly disintegrating. Anxious to get to the drill set-up to test our theories, we now turned west toward the narrow outlet of the lake.

We quickly built a raft, which we paddled across to the south shore. Along the north side of the outlet a once mightier flow of water had deposited a sandy ridge, now studded with jack pines. Sand, pines, and a sparse growth of grass was all we saw when we passed there one morning on the way to our raft. When we returned in the evening after a long hot day in the bush, a most magnificent sight greeted us, the whole southern bank of the ridge was covered with vividly colored big anemones.

Here we were, two dirty, hardened, and tough prospectors in spite of the mosquitos and their heavy, rock sample-laden pack boards went down on their knees to pick themselves a posy of flowers, carrying them home like a great treasure.

Our old dock had been lifted and smashed by the ice. When we saw a wide expanse of open water in front of the camp one morning, we decided to build a new, bigger, and stronger one. All day we labored and by nightfall we looked with pride on our construction, "That's a damn good and solid job," we said, but we had been too fast for the forces of nature.

Overnight the wind shifted to the east, grew to a gale, driving the remaining ice, crashing, and groaning towards the outlet of the lake. Our day's work was undone in a few minutes, torn to bits between the tearing and heaving ice floes.

"She sure is powerful," remarked Steve with a wry grin, looking at the firewood that was our proud achievement of yesterday.

"Yes. 'Idly and with admiration he sees his works perish,' or something like that."

"What you say? Them's not your own words."

"No, just something I remember from school."

For days after the outlet was crammed full of ice. We had to stay in camp, and used our time to repair our canoe with Ambroid glue and canvas patches prior to painting it a bright red. When the wind blew from the west again, once more the left-over ice moved to the east, and our first canoe trip was made along the narrow channel-like outlet. Windy was whimpering pitifully when we left the dock, so we gave in and hoisted her into the canoe for her first boat ride. Somewhat frightened, she lay quietly in the bottom.

"That dog's got more sense than we gave her credit for," I said to Steve.

"Yeah, sure keeps nice and quiet," he called back. Down the calm waters of the channel, we went to where the rapids began, close to the next lake. We drifted over to the south bank. About fifteen feet from land, Windy suddenly got up and made one mighty leap toward terra firma. She didn't quite make it, landed in the drink, and so did we.

Caught completely off guard, we lost our balance, the canoe rolled over, and we were up to our necks in the ice-cold water. A few seconds of a mad scramble and we got ourselves and the canoe to the shore, where Windy welcomed us, putting on a great show of being happy about our survival, or was it as Steve said, "Look at the damned dog, she is laughing at us!"

"Yea, but that will be the last time. From now on she stays home."

"What'll we do now? Go home?"

"By the time we get there we'll be dry. Let's get to a sheltered spot and build a fire, it will keep the flies away, and we can dry our rags."

Shivering, we climbed over the first ridge. Out of the wind we began to feel better, the staccato of our chattering teeth growing softer. Thanks to the water-proofed matches and birch bark, we soon had a roaring fire going, its heat and the sunshine penetrating our bare hide and drying our clothes.

Continuing our march, we kept looking for the westward extension of the mineralized belt. We found it too, a much-shattered zone of deeply disintegrated rock on the surface, heavily stained with the bluish green stains of weathered copper minerals. Following it to the west it led us to the shores of the next unnamed lake, to the east it ended as expected, north of the original find, and with those facts established, one theory was proven. If we needed any further proof, we got it when we climbed the big, almost barren granite ridge, south of the first drill set-up. This gave us a clear bird's eye view of the displacement caused by the younger intrusive rock. There was a certain satisfaction in establishing these facts, but with it there also came a feeling of regret that all the drilling had been wasted. A little more cracking of rock, which to my mind is half of the prospecting game, might have assured success.

With several weeks yet to go, before we might see a plane again, we had plenty of time to trace the mineral occurrences, follow the established leads, and look for new ones. Again, and again, we came back to the main lead running along the south shore. Here and there broken by indentations of the shoreline, it continued past the easternmost end of the lake, and finally disappeared under the overburden of a deep draw.

Hundreds of samples of well mineralized rock were brought back to camp from our daily excursions, all numbered and clearly marked as to the location they came from. A log was kept of these samples, and their numbers marked on an enlarged sketch of the territory surrounding our lake. This was systematic prospecting. Our main task was thus accomplished, but day after day we went farther afield, urged on by the idea ever present in a prospector's mind, *Maybe the next ridge or the next valley hides the biggest find yet, free gold, nuggets big enough to carve your initials on.* It was especially the country around the east end of the lake which intrigued us. Many sheer zones and deeply weathered rock made it look promising.

Often the lake was rough, and we had to load rock ballast into our canoe to make travelling easier and safer.

One evening, returning to camp, we had to battle the whitecaps for hours, sneaking along the shoreline. A slow and tiresome business at the end of a day spent in climbing over rocky ridges and battling through the undergrowth of the draws. Harking back to long-gone days, when sculling and rowing were my sport and pastime, I suggested that we fashion some oarlocks, oars, a seat fastened to the middle of the canoe and a rudder. A rainy day gave us a good chance to try the scheme. It was a great success. With both of us sitting much lower than before, we had a very stable craft, one man sculling with the long oars gave us more speed than two men could achieve paddling. It was now easy to keep on the course, even if we didn't hit the waves at a right angle, and we could turn on the spot.

"All we have to do now is to close in the front end with a piece of canvas. Then let 'er blow," said Steve.

After our chilling bath we had been trying to keep Windy in camp, partly because we were not anxious for a repetition, and partly because we were still hoping to meet up with a caribou. Fresh meat was only a memory, and we were hungry for it. For a few days we succeeded in keeping her in the cookhouse, with a stick pushed into the thumb-latch. Then, one day, deep in the bush we heard faint noises, coming towards us.

Steve stopped and cocked his head, "Listen, something over there! Maybe caribou."

Quietly we waited, wishing we had our rifle with us. Too far to get it from the canoe. Nearer comes the faint rustling in the undergrowth, and then hardly trusting our eyes, we see bounding toward us, Windy, barking joyfully.

"'How in hell did that dog find us? We are miles from camp."

"How did she get out in the first place?" All day she dogged our steps. When it was time to go home, Windy was the first in the boat.

"Nothing doing. Out you go. You show us how you got here so fast."

We started out without her. For a few moments Windy stood there on the shore, legs braced, her whole body jerking forward with each bark, vehemently giving us a piece of her mind. Seeing that didn't help, she threw herself into the water and started swimming. At first, we thought she was following us, but not Windy, she was independent of mere humans to show her the way to go home. About a half mile from our starting point two peninsulas, one jutting out from the south, one from the north formed narrows in the lake. Windy headed straight for the tip of the southern peninsula, neatly cutting off the wide bay. There she shook herself, lay down for a while, and then began telling us off again. As that didn't have any results, she crossed the narrows to the northern peninsula.

"Now she'll go overland." we said, by now really admiring her performance.

"No, sir, look at that dog. I'll be damned. She's cutting off all the bays!" So it was, always heading for the next point, Windy saved herself a lot of travelling over rough, rocky ground, and only minutes after we docked, she made camp too. The cook-house door was ajar, the stick lying on the ground. Windy must have worried the thumb latch from the inside until it had slipped out.

"No use trying to keep a good dog down. Come over here Windy, special treat tonight, a whole can of sausages," Steve's admiration for the dog was evident.

A few gulps. Bark, bark!

"What's the matter now?" Bark, bark.

"What do you want? All gone. Holy mackinaw!" Bark, bark again. Enamel plate is pushed around, a few more licks, and then a soulful look. What's a guy going to do?

"OK. We will open another can. Take it easy now!"

With the end of the quiet and plane-less break-up period, we expected the idyllic, if somewhat lonesome life at Regent Lake to come to an end too, and I personally hoped that the first plane would bring me the long-awaited news that Ottawa would give the OK for my fiancée to come to Canada.

For almost eight weeks I had had no more news from the outside, no mail had reached us, and we had only seen one other human being. We even had to run after him to get a good look.

One morning, Windy signaled that something unusual was going on in our neighborhood. We heard her furious barking being answered faintly from the other side of the lake. A little later we saw a lonely canoeist making his way toward the portage to Frigate Lake. We hailed him, but he paid no attention. Was this another slightly bushed Per? Our curiosity aroused, we got into our boat, and went after the elusive fellow. On the far side of the portage, we found his canoe, but only after at least an hour's waiting did the owner show up.

A skinny, black bearded fellow with wild, restless eyes, surrounded by a bunch of hungry looking dogs, a rusty .22 rifle in the crook of his arm, he came close to where we were sitting, sat down, grunted something which might have been a greeting, stole a few furtive glances in our direction, and then stared across the lake. After a long while the stranger suddenly raised his rifle and fired two shots at a lonely duck swimming way out in the lake. He missed, which wasn't surprising at that distance, and muttered, "Black duck, no good anyway."

The silent session had lasted long enough for us, so I said, "Look, we are going back to camp, why don't you come and have lunch with us?"

"Oh, no. I've got lots to eat. Thanks."

"Ah, come on. We are just as lonely as you are. Change will do you good." With that we left to give him a chance to make up his mind. A slightly eerie feeling it gave us, to turn our backs to the silent one, after all, how crazy. He looked wild enough. Not until we reached our canoe, did we hear him come along the trail.

"Coming? OK. Get in the middle."

Steve and I prepared a good meal, our visitor silently watching. He sure could get along without much talk, but boy! Could he eat!

Who he was, where he came from, where he was going, we didn't find out, not even his name.

Soon after the meal, he asked us to take him across the lake again. At the portage he jumped out, murmured "Thanks," and ran off as if the Devil were behind him. A strange, lonely, and forlorn wanderer in the northern bush.

Sometimes far, sometimes near, the drone of airplane motors was now in the air and finally Pat showed up. "How goes the battle?" the familiar question again. "You fellows find anything?"

"Lots. Come and have a look!"

"Holy Moses. Where did all this stuff come from?" Pat was impressed.

"From the lead we missed. It's quite a show, isn't it? We traced it from the east end of the lake to the next one. Here is the map with all these samples marked in. She's all yours."

"Did you bring any mail?" I wondered.

"Yes. Here you are. Nothing from Ottawa, though. I've written to my old man again. I don't think it will be long now."

"I sure hope it will work. It's been a long wait."

"Have patience, man. You'll be hitched for the rest of your days."

"Sounds good. What happens next?" I asked.

"Do you want to stake any more claims here? If so, I'll send you a few fellows to help. That will be all we'll do here, I think. You will have to build a place at Moore Lake for your honeymoon."

"I had planned on spending it here, but if the work is finished, I suppose that will be the best thing to do. Too bad these good solid cabins will soon be deserted. However, we may come back, who knows? Staking a few more claims would be a good idea, to cover the eastern part of the lead. But for heaven's sake, send me a few experienced fellows. Had a hell of a time working at Hopeless Lake, some of the guys got themselves lost."

"All right, I'll arrange that. Anything else?"

"No, but you haven't told me yet, what kind of a good time you had on your trip. Apparently, you didn't get to the South Seas. I didn't get a postcard from Hawaii."

"That's right. I had to go east to get a few things settled. Had a good time on the West Coast though before that. You know Woods got himself hitched."

"Yea. I heard about that. When are Ronny and I getting our shares?"

"Holy cats! That again."

"Sure. A bargain is a bargain. We fulfilled our part. There's no way out. We have your word on that deal, so you'd better get things straightened out. As far as I can see, your favoured friends crossed you up once more."

"How's that?" Pat looked puzzled.

"You know damn well they peddled off the shares in Edmonton, contrary to the agreement that we wouldn't sell separately. Lot of good these agreements turn out to be."

"How did you hear about that? I only found out after it was done, I assure you."

"Now, look Pat. I have plenty of patience, and I've been a sucker many a time, I don't mind admitting, but in this case, there will be no giving in. You got yourself into this mess, now get yourself out of it again."

"That will do, I mean to say … God bless my teeth … I listen to more stuff from you than from anybody else. Why? I'd like to know myself."

"You know very well. But you won't admit it, not even to yourself. I remember a fellow who used to declare that the last thing he wanted in his outfit was a bunch of yellow livered yes-men. There was to be frankness and straightforwardness in all our dealings. We, the original bunch were going to do this, we were going to do it right. Shall I continue? You know what I mean. We are beginning to lose out, we've lost a lot already."

"Maybe you are right. I didn't like the looks of the main camp when I got back. However, I shall see you there in a few days. We'll try to get things straightened out. So long then, Hermann."

"So long, Pat. Don't let the Irish in you get you into more trouble."

"I am in it now to my ears. You don't know the half of it. God bless Ireland. So long."

"Seems she needs it. So long."

Ronny and Punk reached our camp a few days after Pat's visit to help us stake the claims.

Punk was full of stories about Moore Lake.

"You guys sure have it nice and peaceful here. Damn sight better than the other camp. Guys are all going nuts there. Pat brought a dame into camp; she's got them all going. Can't blame the guys, most of them had almost forgotten what a skirt looks like. Now they know, and don't know what to do about it. Gripes, it's tough, I tell ye."

"How about you, Punk? Having trouble too?"

"No, siree, this here Punk is going to help himself pretty soon to a real nice trip. I've got a pretty fair stake. There will be the bright lights, nice cool beer parlors, and some steaks smothered in mushrooms at Johnson's. About the rest we'll see later."

"I see Art gave you the right instructions how to enjoy life in Edmonton. Well, send us a postcard when you get there."

"Don't count on that. Little Punk will be awfully busy. What with dodging cars and street cars. What are we going to do here?"

"Stake some claims at the east end," Steve explained.

"For the love of Pete! That again. Pretty soon they will have the whole goddam country staked from Athabasca to the North Pole."

Ronny, quiet as usual, didn't say much to all this. Women didn't bother him yet, to all appearances, or was it that he didn't believe in a lot of talk? He was a fellow for action, all the words and tall stories were to him just a poor substitute. Slightly, amused, he watched the antics of his chums, once in a while slamming on the brakes, when stories grew too tall, with a plain, definite statement, "I don't believe that."

While the tall stories were being spun, we sharpened our axes for the staking: a job that didn't take us long with four old-timers on the go. Following that, Steve and I took the other two fellows around to show them the mineralized veins and outcroppings we had located. A pleasant and interesting week was spent, untroubled, carefree, only of myself I have to say, that again and again the question arose in my mind: Will the final word from the Government come soon, will it reach me in time to allow me to go to New York to meet my fiancée as we had planned? But there was no official-looking letter in the mail that Pat brought to camp. His arrival, however, meant goodbye to Regent Lake for me. It was time to build a little cabin for two.

Somewhat sadly I packed my belongings, and had a last look around the camp, where, although the drilling results had been disappointing, I had lived many happy days in the cheerful company of a splendid bunch of fellows, days that I knew I would always recall with deep pleasure.

CHAPTER XXV
SUMMERTIME

IT WAS ONLY a short hop to Moore Lake, but it almost seemed like a different world. Calling on Maud and seeing Little Pat, sturdy-legged, rosy-cheeked, and full of life was indeed pleasant and reassuring.

Walking through the camp right after landing had been somewhat of a shock. I missed the orderliness, the dock was cluttered up with a great collection of material lying around helter skelter, and only after we had tied up the plane, a few fellows, new faces mostly, came in sight, as it then seemed with rather surly expressions on their mugs. Here indeed was a great change, and it all looked like a confirmation of Punk's report.

A little later I saw one of the main disturbing elements saunter down the path between the main camp and Pat's cabin. Auburn locks falling almost to where the slacks, on the tight side, started, provocatively swiveling hips, no wonder three dozen healthy young fellows all thinking themselves as the perfect answer to a maiden's prayer were in ornery and scrappy mood.

As old Alex put it, "Maybe I'd better start putting some saltpeter in the grub, or our young bucks will get out of hand. As long as they don't see an unattached female strutting her stuff before them, they are quite all right, they dream of what they are going to do when they go outside with their stake and they talk about it. Sort of a safety valve. But this is tough on them." To me, the whole affair looked like the deplorable outcome of a noble gesture on Pat's part. The inherent Irish sentimentality must have gotten the better of sound reason. However, this was none of my business exactly, and as this episode was bound to come to an early end by itself, it seemed then, and it seems now, best to ignore it.

As I tried to recall my years in the North, I could not escape the conclusion that the unpleasant things were caused by influences that had their roots outside of the northern bush. People who invaded the North, lured by quick riches, with only one thought in their minds: to make a big stake and get out again, they were the ones who brought the trouble.

Fortunately, in the long run the North weeded them out. Only those whose heart and mind could find peace, contentment, and delight in the everlasting beauty of unspoiled nature, and who enjoyed pitting their own strength against hardships and harshness of the elements, and feeling that strength grow with the greater demands remained. They would not be the ones who told the fantastic stories, to impress the outsider. With a grin, sometimes they told a true, but daring yarn, but it was more likely that they would be quiet guests with a look in their eyes that seemed to scan the far horizon, where the end of a rainbow dipped into a silvery northern lake.

Working away at the first home my future wife and I were to share, I was looking ahead impatiently to the day when together we should wander through the bush, when I could say,

"Let your own eyes tell you, and listen to your heart, and you too will feel what with my own feeble words I tried to convey."

Urged on by the hope that soon the final word would come, I labored long and late. A floor was put down, a frame erected with two-foot boardwalls, and over this a 9 x 12 tent was stretched. Solidly built into the framework were the window and doors, one of the latter a French door made from window sashes. The walls were doubled on the inside with building paper and canvas. A corner cupboard, chairs, a table, bookshelves and shelves for the kitchen utensils, a small cook-stove and a bed, all furniture hand-made, didn't leave much space to move around, but it presented a cozy picture, and after all, there were the great outdoors right at the doorstep, if claustrophobia should set in.

Everything was shipshape to welcome the bride, and still no news.

So, I was stuck, and time grew long. After the hectic days of building, I now didn't quite know what to do with myself, and in a way welcomed Pat's suggestion to ease the time of waiting by doing some prospecting around a lake we had often flown over on our trips to Goldfields.

"As long as you get word to me immediately after the official OK arrives, I'd rather spend the remaining time in the bush. I am not very cheerful company for anybody these days, I'm afraid, but I believe Steve will put up with me. But, for the love of Pete, make sure there is no delay," I explained that I did not want to just hang around waiting.

Pat agreed, "Don't worry. By now we are just as anxious to get you hitched as you are yourself. I'll get Steve from the other camp, so you can start tomorrow."

Mac Lake was an almost square body of water, hemmed in on three sides by rocky ridges, while the south or east side was a lovely long beach of fine sand. Here we built our temporary camp and started the old game of looking for mineral.

A great number of quartz veins cutting through the country rock, some of the bigger ones we had seen from the air, indicated great action in long past days, but the mineral wasn't there, the quartz was white and sugar like. Wider and wider circles we drew in our daily wanderings through the bush, and I kept my ears cocked for the sound of an airplane landing near our camp. After about a week Pat appeared. Full of hope we ran to the plane, aground on the sand about 50 feet from shore. No news for me.

News of another kind. Some of the fellows at Yellowknife wanted to go outside. To replace them, a lot of staking still had to be done down there, Steve and some others were to go north.

"I really thought you had news for me, Pat."

"Yea, too bad. Maybe I'd better wire my old man again. Don't worry though, by the time we are through, he will have not only Ottawa, but the CN and CPR going, I'll bet you. Keep your nose dry."

My new companion was a friendly little fellow, but an absolute beginner as far as life in the bush was concerned. He gave me some anxious moments, lost on our first trip. I finally found him; he was almost in tears. How he got lost, I don't know, but he never took a chance again, never let me out of his sight. Days wore on and became weeks.

Being so close to Goldfields, the sound of planes could often be heard, and added to the state of restlessness and expectancy. So, I suppose I finally had the proper bridegroom-to-be jitters, because in retrospect the events of these anxious days seem somewhat beclouded.

After many more long days finally there arrived a Canadian Airways plane with the decisive news of her immigration approval, which had grown slightly old on their way to me. It was quite possible that the news might have reached my fiancée in Germany before I received it in the North.

This was a fix, and my humor wasn't improved by the pilot's statement that he couldn't take me right away but would pick me up in the morning. "Pat told me to get this to you immediately. That's all I could promise because I have two more charter flights today. Don't worry! Tomorrow morning, for sure, I'll come and get you."

CHAPTER XXVI
WEDDING TRIP

BAD WEATHER INTERVENED. Only three days later I finally got back to Moore Lake and from there to Goldfields, where I was lucky enough to catch North Saul who getting ready to fly south via Fort Chipewyan, in his old, slow Fairchild. Poor North wasn't feeling any too good that day. I found him sitting on one of the floats, holding his ailing stomach.

"Will we make Edmonton?" I asked him.

"Don't know. We'll try. I'd like to get there myself; I feel like hell. But I just got word to take the Indigenous Chief to Cannery Bay and pick up some stuff there. Here he comes now. Let's get going."

A long stop at Cannery Bay, then a longer one at Fort Chipewyan, and our hopes to make Edmonton grew dimmer. A heavy load, a strong headwind, and evening began to fall before we landed at Fort McMurray.

"Do you think we'll make it, North?" I asked.

North patiently put up with my eternal questions, although he had his own troubles. "I have my doubts," he replied, "we'll try, but if it gets too dark, we'll have to stay overnight at La Biche."

There were some surprised faces down at the seaplane base when we cast loose once more. North steered his plane toward the south, twilight was creeping over the country, and soon we had to realize that we would be lucky if we got safely halfway to Edmonton. Darkness was spreading rapidly when we slid down to the placid waters of Lac La Biche.

"Too bad," said North, "but if it will cheer you up, you can get the best feed of fresh pickerel at the hotel."

"Pickerel? Oh, yes, I missed out on that when I came through here in 1935."

We had a good meal and turned in early to be ready for the last lap at daybreak. A light mist was rising from the glassy waters of the lake, when we got down to the dock, as the first rays of the rising sun gilded the scene. An hour's flight and Cooking Lake was below us. Summer heat was brooding over the flying base, and we were glad to find a Canadian Airways car ready to take us to Edmonton. The next stop was the King Edward Hotel for a hurried bath, a change to civilized garb, and I headed for the steamship office to find out when the *Deutschland* would dock in New York. I was too late, there was no way of getting there on time.

A solution to my troubles, I learned to say "our," could have been found. I could have turned to Uncle Henry and Dorothy to take care of my beloved until I got there, but the best thoughts, alas, came too late. I plead guilty to not having had all my wits about me.

While I waited in Edmonton, the days of waiting grew long, although, with the help and advice of Wop, I made arrangements for a small wedding party at the Springer Hotel.

So came at last the great day, when the westbound train with its precious passenger was due. Four of us, Maud, Pat, Wop, and myself made up the welcoming party, and though the days before that great moment were be hazy in recollection, the picture of the one who had filled my hopes and dreams, stepping off the train and into my arms, is clear and unforgettable, and cherished forever in my heart.

Time now fled as never before, so much to be told, so many friends to be introduced, many a toast to be drunk.

We almost missed the Justice of the Peace, who was ready to go home, when we appeared on the scene to get the official approval. We were married!

The wedding came off, and it came off well, thanks to the spunk of the girl I had chosen, who gave up her sheltered home, crossed the Atlantic, and now the wide stretches of Canada, to come to me, to begin life together. Life in the North, as we then thought.

A few more toasts to health and happiness at the King Edward, and more of them at our wedding dinner, and when the wine was drunk and the candles burnt lower, Wop, with almost fatherly solicitude, took us in his car for a long ride through the moonlit, peaceful, balmy midsummer night, along the Saskatchewan River. Through a lovely, enchanted night of deep purple shadows, a silvery moon and dreams fulfilled.

Gay and shining summer days followed. On the go all day, there were sights to be shown, more friends to be met, and life together grew into a reality, keeping dreamlike charms.

"Happy? Glad you came? Like the Canadians?" Again, and again, I turned to my bride.

Strong and joyous came her answer, "Oh, yes! I am glad and I like your friends. They are so friendly, and they seem so happy, always ready to smile and laugh. I think that is wonderful. On the other side, you know, people have forgotten how to laugh, and even their smiles are tinged with bitterness, but I don't want to think about that, not now. I am so happy to be here with you."

"You made me happier than I ever dared to hope for, Dearest. We will keep it this way, won't we?"

"Yes, always!"

"Always."

CHAPTER XXVII
A NEW LIFE

MANY FRIENDS ASKED us, "Where are you going for your honeymoon?"

"Down North" was our answer, and so one morning we found ourselves at Cooking Lake, ready to board a Norseman piloted by Paul Davoud. Paul flew for the Royal Air Force before joining Canadian Airways Ltd, so we were in good hands. There were six passengers for the northbound flight. Maud and Pat, my wife and I, and one lady photographer, bound for Fort Chipewyan to take pictures of wild fowl in the Athabasca delta, decorated with an impressive array of business-like cameras and an everlasting smile.

Number six, lo and behold, one lightweight: A canary in his cage, given to Pat by one of his friends. As he said, "If I can believe all you fellows tell me about your comfortable homes in the North, this will give the finishing touch."

With our luggage we made up a good payload for the Norseman. For the take-off, all crowded toward the pilot's compartment, but once airborne, we settled down to the long flight as comfortably as was possible amidst suitcases, duffel bags, boxes, and crates.

The Norseman, in those days one of the fastest planes flying in the North country, made good time. Refueling at Fort McMurray gave us time for a quick lunch at the Franklin Hotel, and soon after we were following the meandering Athabasca. Down below stretched the vastness of the northern wilderness, and what I had felt on my first flight north I now saw mirrored in the shining, wondering eyes of my new partner who eagerly took in this complete contrast to her former surroundings.

From a crowded hometown where more than a million people live on a few square miles, she had now come to one of the last thinly settled districts that were left on this earth. Conversation was kept at a minimum by the motors' drone, a lot of pointing was done, a few explanatory remarks shouted. Flying for hours, we hardly saw a sign of human activities. Once we had a glimpse of one of the Hudson Bay's diesel boats pushing some scows down river, that was about all. A southbound plane passed nearby, Paul wagged his wings, and on and on we went, finally leaving our good guide, the Athabasca to swing east toward the Williams River, across the big Lake and straight for Moore Lake. The sun was still high in the sky when Paul set his plane down and taxied to the dock. Our long but pleasant honeymoon flight was over.

Quickly we ran to the little "castle in the bush" that I had prepared, a small but friendly abode, and thus began our life together down North. Finally, the long yearned-for had come true, and as we used to say in the old country, "Room is in the smallest cabin for a happily loving couple."

Many things must have been strange for my bride in those first weeks, but with astounding ease she took her place amongst us old timers, despite the fact that not only the surroundings but also the language, were unfamiliar.

But it didn't take her long to establish firm footing in this new world, and the progress towards mastering the language was most amazing. From the beginning we had made a pact to speak only English, and less and less often were explanations in our native tongue called for. This effort to speak English fluently was unknowingly spurred on by the ever-teasing Pat, who, by and by, got paid back in the same coin he was dealing out; some great verbal duels took place in our evening meetings.

During the day, whenever there was time, Lotti and I explored the surrounding bush or went for canoe rides on the lake. One sunny day followed the other, each one filled with delightful new experiences, but it was our first hike through the bush, to the north from our camp, which remained most deeply engraved in our memories.

It was toward evening, when we made our way along an old caribou trail across a thickly wooded range rising not far inland from Moore Lake. The air was calm, the forest was drenched with silence, only occasionally broken by the cracking of a dry branch under our feet. Once over the crest of the range we made our careful way through a thick patch of undergrowth, and then with dramatic suddenness we found ourselves at the shore of the loveliest lake. I had seen it before but had not given a hint so as not to spoil the surprise, but even to me it seemed more enchanting that evening.

Graceful birches and somber evergreens growing along the shoreline were mirrored to perfection in the water which seemed like a highly polished sheet of silver, here and there gilded by the rays of the westering sun. In silent admiration we stood and let our eyes feast, and the peace of the quiescent scene before us filled our souls. As we thus stood and the sun sank still lower, a gossamer violet-blue haze seemed to flow into the atmosphere, softening the brilliance of a moment ago.

"Time to go, Dearest. Night is falling."

"Yes. But we must come back to this, to our lake. I cannot say it, I have no words. You know."

"I do, and now we both know."

September 1937 came. The nights grew longer and cooler, reminding us that soon Winter would hold sway again, and we had better be prepared. No word had as yet reached us from our company about further work, but we were filled with confidence in the future. Thus, with enthusiasm we began the construction of our winterproof big cabin.

Hardly had the walls been raised when the blow fell. Pat had been to Goldfields for the mail. He returned with a deep scowl on his face that boded no good.

"What's wrong now?" I asked him.

"Bad news," he replied, after he clambered out of the cockpit. "Come along, I'll show you."

Away from the dock he held out a wire to me, "From the big boss, read it."

There it was "no work for the winter".

We had arrived at his cabin, somewhat dejectedly we sat down on the verandah overlooking the lake and talked over the implications of this unexpected turn. "It's tough, Hermann, tough enough for us, but what will Lotti say?"

"She will take it, just as well as we will manage to take it. Here she comes. You will see."

My smiling, happy companion appeared, looked at us: "What's the matter? You both look sad!"

"We had bad news. No work for the winter."

"Oh, is that so bad? If we are sad, it will be worse!"

"Well, for the love of Pete. But she is right," Pat exclaimed. "This calls for a drink," and he went to get the wherewithal. Maud returned with him. Pat, crossing glass and bottle, "Help yourself," made the round. And so, we raised our glasses to the future that looked particularly dim. We were ready to face it.

CHAPTER XXVIII
LIGHTS AND SHADOWS

WITH MORE ATTENTION than ever we listened to the news on the radio. We knew that the world political situation was deteriorating more and more, one crisis seemed to follow the other, and no doubt the reaction on the stock market had in turn led to the bad news Pat had received. About all these things we talked in the evenings, in our minds revolting against the unhappy fact that if a few stock market gamblers got jittery we should feel the repercussions of a market slump deep in the northern bush.

Realizing full well that the mining industry would be the main factor in opening up the North, we saw at the same time the weakness of this set-up. Instead of a sound, steady development, which we considered a national task of the greatest necessity, there would be sporadic rushes and lulls, caused by the fluctuations on the market thousands of miles away. These eternal short-lived ups and downs, often brought about by the manipulations of speculative elements, are definitely contrary to sound long-range planning, that alone would unlock the North's treasures for the benefit of those who rightfully should be the heirs to them, the Canadian people.

Often, we had talked about all these things. That our ideas were sound had now been driven home very forcefully. Here we were in the bush. We liked our occupation. In a short while we had shown that we were able to find the stuff. We wanted to make it our life's work, to go on and on, and with increasing knowledge and experience do our share in opening up the North.

This was our ambition, definitely a worthwhile goal. But we were told, "No work for the winter." Would there be work in the Spring? We didn't know. We knew one thing though; The North was still waiting!

Long discussions about the next steps to be taken, led to the decision to stay at Moore Lake over the winter. To find a short-term job in Edmonton would be difficult. Nobody had any intention to leave the North for good, and considering all angles, nowhere could we hope to live more economically. Once we had decided on this, our spirits rose, we had been rocked back onto our heels, but we weren't knocked down yet. With new vigor and determination Lotti and I finished our cabin. Work was the best chaser of gloomy thoughts, and as new ideas concerning the interior finishing constantly popped up, there was plenty of work.

We were happy when finally, we could move into our cozy cabin. With well chinked log walls, a jolly red and green trim around windows, doors and along the eaves, it looked very inviting. The interior fulfilled the promises the friendly outside made. Let winter come, we thought, it will find us well prepared.

Confidently we looked at our four-foot camp-stove, a very capable looking affair. We had used the same type at Regent Lake. Before retiring for the night, we used to put on some green birch logs. With draft slots and key damper closed tight, they smoldered all night. In the morning we had only to open the draft and the condensed tars blazed up into a roaring fire. Rattling, puffing, and huffing, little tongues of flames shooting out of the draft slots, the old stove worked itself into a red-hot rage, and we could crawl unhesitatingly out of our bedrolls.

Colourful curtains were put up as Lotti soon discovered the magic of the mail order catalogue so rugs, pictures and all the other feminine touches, which make a house a home, softened and brightened the hand-made furniture. Great satisfaction and joy filled our hearts as we looked at this our first real home, which we had built ourselves.

According to further information received at that time, our company rather favored our plan to stay over winter at Moore Lake. A great deal of equipment and large stores of supplies would thus be guarded and preserved for future use. However, we would not be on the payroll, and although this was a set-back, we figured that things would be more difficult in the city.

The days, growing ever shorter, were still warm and sunny, but the long nights grew colder. There were other signs. The geese were flying south, sometimes heard and often seen, mares' tails in the sky, and in the mornings, milky white mists rose from the lake. The leaves of the birches and poplars suddenly turned yellow and golden, and soon began to fall.

At first it looked as if about ten to twelve people were going to make up the camps' community, but when freeze up came nearer, Maud and Pat suddenly decided to leave for Edmonton, for a short time only, we were given to understand. However, they didn't return. The reasons were their own, but that sudden decision had a very upsetting influence on the original plans. For Lotti and myself, it was a great disappointment, not only would we miss their company during the long winter evenings, but it seemed then, and this unfortunately came true, that with the leaving of the field manager, some might act in a manner that would make cooperation in the common interest very difficult.

I wanted to tell of the North. The disturbances in the human relationships as they arose during the fall of 1937 in our camp, might have happened anywhere else, and as mentioned once before, they could be led back to outside influences. This is the reason for dealing lightly with those affairs, and not to save somebody's feelings. Ours weren't considered very much in this emotional turmoil.

As members of a small, isolated group, it wasn't easy, even with the best intentions to keep out of the trouble others insisted on cooking up, but it would have been less wearing had we stayed in our cabin, instead of moving into Maud's and Pat's log mansion, at their suggestion. Looking back, the reasons for that move do not seem strong. It was a big mistake; we learned the hard way. Judging these happenings with time-grown detachment, I believe it was in a way a repetition of the dirty maneuvers which led to Bob's leaving us at the time I returned from Europe.

Pat's absence was used then to make things unpleasant for Bob, who was held in high esteem by the former. Greed and jealousy were the motives of those who managed to discredit him. This, and the only too human inclination to distract attention from their own troubles by making trouble for others, led to the difficulties that marred the days at Moore Lake in the early winter.

Woods, who must have lured his spouse into marrying him by golden promises of quick riches, had seen his dreams shattered on the hard rocks of reality. The North was too much for his better-half, especially as the gold seemed somewhat far off just then, and poor Woods, at his wits' end, became moodier every day, lamenting with gruesome monotony, completely off key, "Out of a blue sky a dark cloud came rolling!" Day after day, until it got on our nerves, like Friday's circus song of long ago.

All this wouldn't have fizzed on me, and the brave trouble-cooks, realizing this, therefore decided to get at me by making things unpleasant for Lotti, who was rather defenseless because she was not yet familiar enough with the English language.

Shortly before freeze-up we finally heard some cheering news again. North Saul, the pilot, who regularly dropped in at our camp and brought in our mail told us that he would stay in the North over freeze-up. If possible, he would like to stay at our camp and change his machine over to skis with our assistance. We warmly welcomed this news. We liked North, a quiet, serious fellow, he would be good company, and at the same time we hoped that his presence would put a damper on the caprices of certain troublesome persons.

When North flew his old Bellanca home to Moore Lake from his last flight on floats, snow flurries were dancing in the air. Old Man Winter had come once more. Our hopes, shared by Reg, the plane mechanic, and Ned, with whom we got along most pleasantly, for a peaceful interlude, happily came true. Long hikes through the silent, snow-enshrouded bush, followed by hearty meals and lazy sessions around the fireplace were enjoyed to the fullest.

No flying messenger would reach us now for weeks, but Reg kept the radio sending and receiving set tuned to B&M's station on a regular schedule. Religiously he would come to the cabin on time, stamp the snow off his feet outside, step into the door quickly, rub his hands and declare, "Boy! She is a cold one today!" With great patience he tuned the set, resulting in fearful howls and hisses. "Boy! She is coming in strong tonight!" and then, usually silence.

Not very often did he achieve good reception, but somehow the illusion of still being in touch with the outside world was kept up. Usually, we were glad to turn our broadcast set on again or go back to our talks or books.

Reg himself was a great talker at times, although it has to be said that he concerned himself chiefly with the sad and messy results of airplane crashes, his versions of the story usually putting the blame on the poor judgment of the hapless pilot.

We had heard all the stories before, but now he had found a new victim in North. It would go somewhat like this, "Ever meet Harry Smith?"

"No."

"Well, he is the guy who pancaked that brand-new Stinson at A. Tried to land in a crosswind."

"Never heard of him."

"Ever meet a fellow by the name of Corby?"

"Can't say that I have."

"Well, he is the guy who overshot the field at B. Hit a power line."

"Don't know him. What happened?"

"Some fireworks! Burnt to a frazzle!"

"Who?" North asked.

"The plane. Corby was thrown clear. Landed on a haystack."

"Good for him!"

"Ever hear of a fellow, Guy… something, I forget. Came from Montreal. He was barnstorming with another guy, what's his name, Compton or something…"

"Yea?"

"You met him?"

"No, never even heard of him."

"Well, you should have seen that. Overshot the field. Knocked the windsock off the hangar."

"That's all?"

"Hell, no. Sheared off the under-carriage, clean as a whistle."

"Oh."

"Yea, plowed right into a barn on his belly."

"You don't say."

"Yep. Dead as a doornail."

"Who? The pilot?"

"No! The cow."

"Holy doodle! I'd better give up flying, seems to be dangerous," North would try to end the conversation.

"Ever meet…"

"No, no, never did and never want to. Seems you met a lot of bum pilots. What about the grease-monkeys? They never pull any boners?"

That was about the only way to stop old Reg. Otherwise, there was no end to his sordid stories. I have cut it short, omitted some of the bloodier details. Some of it was really gruesome and made you wonder how you could ever again screw up enough courage or plain foolhardiness to crawl into an airplane or trust a pilot to get you down safely if he managed to get you up in the first place.

Somehow, though, you couldn't get really mad at the teller of the disastrous stories. He told them all, with such a genuinely sad expression on his rough-hewn features, with the tip of his enormous beak almost touching his nether lip. Maybe life hadn't been very kind so far… I am happy to be able to relate though, that later on, after Reg got himself married and became a proud father, he felt much happier about the whole thing.

Deeper into winter marched the time, daylight came later, and darkness fell sooner with every parting day, and the cold grew more intense. North made his test flight on skis, then took us up for a ride one sunny forenoon. It wasn't only for the ride we went; we were hoping to see some caribou trekking south.

The year before, great herds had passed not far from our camp. This time, alas, not a single one did we espy from high above, that grand shining morning of unlimited visibility. The only sign of life down below was the smoke from our cabins rising straight into the clear still air.

Not long, and North had to return to work. Circling the camp, he wagged his wings in farewell, then headed south where loads of mail and goods were waiting for the northerners. Slowly the short days, and slower yet the long dark afternoons evenings, passed. Keeping the fires going, preparing the meals, and chopping open the waterhole every morning left us with plenty of time on hand. Maybe that was the reason why some began talking about doing some trapping.

I was opposed to this idea, for the simple reason that traps near the camp would endanger our dogs. I thought I did some convincing talking, and when no more was said about it, forgot the matter. Not for long though.

As a rule when darkness began to fall all our dogs would come inside. One evening Ridsy was missing. Ridsy was a fine-looking fawn-colored bitch, a gentle clever dog beloved by all. A strange but strong attachment had grown between her and Buster, a poor, haggard looking frightened mongrel sleigh dog who appeared from nowhere in our camp one day. Soon it became evident that there would be consequences, and we thought the day was near that the puppies would arrive. Now, Ridsy was gone. Did she go off to have her pups in some quiet corner all by herself? When daylight came, a thorough search was made in and around the camp. No Ridsy. All day we searched and called, and all of the following day. Not a sign was found, it seemed hopeless.

Then, three days after Ridsy's disappearance, Reg came running from the cookhouse one night.

"Ridsy's at the cook-house, all played out. She's hurt too. Come over quick!"

There she was, more dead than alive, her eyes filmed over, her nose hot and dry, and a fierce-looking festering gash across one front paw.

"Some damned fool set some traps after all."

The damned fool finally admitted sheepishly, "Yes, but a long way from camp."

"So much worse for this poor dog. No time to talk about this now. Let's try and get some hot mush into her and clean the wound."

Near the stove we carefully bedded down the poor shivering creature, faintly hoped for a good outcome, and returned to our cabin. There was no great change the follow morning, and so we were not in the least prepared when we heard the scratching at our cabin door early in the afternoon.

As soon as we opened the door, Ridsy came limping in and made straight for the child's bedroom. "Can you beat that? She knows the right place." A bed was made for our patient. Lotti took over the nursing, and a hopeless task it seemed for many days. The crisis came one night. We thought it was the end and sat up way into the small hours of the morning until at last Ridsy's labored breathing grew quieter.

Small whimpering sounds wakened us from our slumber. Hurriedly we dressed and rushed to Ridsy's bed. She was just giving birth to a third pup, even feebly trying to lick her offspring dry. It touched us deeply to see her try so hard. But it was too much, completely played out, she let her head drop and looked at us, so it seemed, as if she wanted to say, "You take over, it's too much for me."

We had our hands full, what with starting the stove, heating some milk, and drying puppies. Reg then came in near noontime for his radio schedule.

"Boy, she is a cold one today! What's going on here?"

"Ridsy is having her pups!"

"How many?"

"Three so far. Have a look!"

"Boy, they are nice. Have to tell the others." Off he went, soon to return with Dan.

"How many?"

"Four. That seems to be all for now."

After a long interval number five appeared. If we thought that would be the last one, we were very much mistaken. The pups were all lively enough, but with each new life coming out of Ridsy, the poor creature seemed to give out the rest of her strength. When evening came, we had eight lively puppies, all healthy and well built, but their mother looked on the verge of giving up. However, all survived, thanks to the good care given them by Lotti who spent hour after hour feeding the weak mother and her lively offspring. Lotti's maternal instinct was developing.

It was a busy time, and rather peaceful around the camp, which gave us hope that Christmas might pass without the constant rumpus certain people in camp kicked up with monotonous regularity. Thus, with eagerness, preparations were undertaken, and with greater eagerness were the mail bags opened, when planes began to come in regularly again.

One day, Doc from Goldfields crawled out from between the boxes, crates, and mail bags, to see how we were. When Lotti and I were alone with him, he took on a very fatherly air, "In my opinion, as you are pregnant, you should go to Edmonton, and not take the chances that you might find yourselves without medical help. Everything is fine so far, but with the first baby coming, one has to be careful." Doc was right, we realized that, but we hated the idea of being separated. Sadly, it would be a long separation as the baby was due around May.

"After Christmas be all right, Doc?"

"Oh, yes. There is no hurry. But as I don't know when I can come back, I wanted to tell you now."

"OK Doc. We'll spend Christmas here. After all, it's our first one together."

Only a few miles away from us old friend Pete Lauder was spending the winter by himself in the smallest log cabin we had ever seen, about 6 ft. x 8 ft. With the walls deeply banked with sod and moss, it looked even smaller. Under the snow it would be more like a dugout. Many years in the bush, trapping and prospecting, no doubt, had taught Pete how to make himself comfortable without heating up a lot of air. "As soon as I can cross the ice, I'll come and see you," Pete had said, when we last had seen him just before the onset of winter. Here it was going toward Christmas, and no Pete had shown up. This began to make us wonder, although we knew Pete had spent many a winter trapping alone in the bush.

So, one morning, a couple of fellows set out to check up on the old-timer and found him well and quite surprised that anyone should worry about him. The trapping had been very poor, and what was worse, a big herd of caribou had passed right next to his cabin while he was inspecting one of his traplines. His food getting low, he would soon have come to see us.

"Would he come to spend Christmas with us?" Oh, that, he couldn't say yet, "Maybe."

"He wouldn't say yes?" we asked the boys on their return

"No, nothing doing, wouldn't commit himself," the boys had tried.

Old Pete was running true to form, however, I felt sure we would have a guest, loneliness can be borne but it's harder at Christmas… Many a time we looked across the lake during the short daylight hours on December 24th, only when darkness began to fall, did we finally see a small dim shadow grow into the figure of a man on snowshoes, and then old Pete was with us, resplendent for the occasion in a brand new brightly colored shirt, new overalls, and new mukluks.

We were all dressed up in a more city like manner, but big, barrel-chested Pete cut the most impressive figure. By the cozy light of a birch log fire and the gaily decorated table, in almost forgotten harmony, we sat down to our Christmas dinner.

We were happy and content, but the happiest of all was our guest. There was an old worldly charm about his perfect manners which captivated and intrigued us all. Just the way he always turned towards whomever he was talking to, leaning slightly forward, giving his full attention. The way he held his cigarette … it was striking, more so because it seemed so genuine and absolutely unaffected. Pete definitely was a man of the world, and later, many times, we wondered where he had come from and what made him go forth to trap and prospect in the loneliness of the Northern bush…

The pleasant hours of the holidays passed only too quickly, and partly because Doc had said so, partly because of beginners' jitteriness, we now had to face it, another separation. The baby was expected in May and it was time for Lotti to head for civilization.

With heavy hearts we decided to make arrangements for Lotti's flight to Edmonton. Reg managed to get the message through to Goldfields, which brought Con Farrell, one of the ace bush pilots, breezing into our camp.

"I'll pick you up on the 31st early in the morning, so we'll be in Edmonton for New Year's Eve." That was Con's plan. But, as I later learned, the arrival of the New Year had to be celebrated at the Hudson's Bay Post in Fort Chipewyan. Lotti didn't sleep much in her room above the bar as the New Year's Eve celebration below was somewhat noisy!

Heavy fog over the lake and the Athabasca Delta made the continuation of the flight to Edmonton impossible. Had we only known! We might have toasted to the New Year together at Moore Lake. A forlorn threesome had the snow blown into their faces, as Con gave his motor the gun and pulled away into the wind. He had a full load.

At the last moment most of the leftover inhabitants of our camp had decided to pull out, leaving just Reg, Dan, and myself behind. Off and on it had been hinted that Pat would return to camp, and by and by we had given up hope of this coming true. A visit by the big boss hadn't given us any encouragement concerning our chances to see work resumed in the Goldfields' district the coming spring. I had hopes that Lotti might be able to get a clearer picture from Pat once she got to Edmonton.

The separation went cruelly against our plans, and feelings, and for this reason we decided beforehand that, unless we soon learned something positive about the future of our camp, I would follow to Edmonton, so that at least until springtime we might be together. To leave the dearest girl in the world all alone in the city, where she to hardly knew anybody, having to express herself in a still unfamiliar language, that thought was driving me into a terrible restlessness during the ensuing days and weeks, and finally became unbearable.

With the three of us batching it in great harmony, there was less work than ever and too many hours left for the mind to wander and speculate.

For the first time I began yearning to get out of the bush, and eagerly listened in when Reg tried to contact the outside world during his radio schedules. I counted more, however, on getting news from Lotti through CJAD's *Hello, the North* program, but as luck would have it, the reception was too poor. At last, it seemed there was only one way out.

Sadly, I went to work in our own cabin to pack all our belongings, feeling that we would not again inhabit the first home we had built with such great hopes and so much love and labor. Cold and barren looked the once so homey place, when I had a last look around, and then I too was ready to go, to leave the place where our hopes had been strongest, and our dreams most beautiful.

CHAPTER XXIX
CITY LIFE

AS A RESULT of the big boss' short and not in any way encouraging visit, a plane came to our camp one day, to carry some of the still large food supplies to the B&M camp. I took this as one more sign that Moore Lake camp would not be reopened in the spring, and when I was packed and ready, I bummed a ride to Goldfields.

Not far from where we landed, I saw a Junkers plane with vapors rising from her nose hangar. I rushed over. There was Payload McMillan, the pilot, looking more enormous than ever in his Parka Hood. "What's the hurry, Hermann, had enough of the bush?"

"Are you going south? I want to go to Edmonton!"

"Well, I am and again I'm not, or am I?" Payload was cryptic.

"I am asking you!"

"Well, fact of the matter is, I should have been gone about an hour ago. There's a sick guy at the Athona. Just when I was going to take off, here they come running, 'Wait for a few minutes, we have a sick fellow to take out.' So, I wait, and wait, and wait. Word is sent: 'the guy is too sick to travel.' So, we warm up the engine again. We are ready to take off. Another message comes, 'The fellow is going out after all, in fifteen minutes.' That's half an hour ago. No patient. Now we are warming up the engine again. "I'll give them another ten minutes. I want to get to Fort McMurray by daylight."

"Here he comes now," I said, pointing to a dog-team racing towards us. There was with it, no sick man, just another message to tell us he wouldn't come.

"What the hell is the matter with you guys? What in hell do you guys think: I am flying around this godforsaken place for the fun of it? What the ... ah, what's the use ... let's get the hell out of here!"

The firepot was turned off, down came the nose hangar, I crawled into the capacious belly of the Junkers, stowed away the gear the mechanic handed me. He then cranked the starter. The motor coughed and spat, settled down to an even roar magnified by the resonant all-metal body of the Junkers, and with all horsepower unleashed, we glided faster and faster till gliding almost imperceptibly changed into flight, and we were on our way.

South of the big lake, under a blanket of snow overhung by a grey sky the scenery looked now even more monotonous than in the summertime. Close to the south shore here and there one saw big and small herds of caribou. After that, not a sign of life in the immense white wilderness.

Nearing Fort McMurray, the visibility constantly got poorer. Fog rolled in waves in the country. We just made it. When we crawled out of the plane even the far bank of the Snye River could barely be seen. That meant a night at the Franklin Hotel, not an unpleasant prospect. There was always interesting company to be found in the well-known caravansary at the jump-off place for the northern hinterland. Hardly had we started to trudge, with somewhat stiff limbs, up to the settlement when we were arrested by faint motor noises pulsating like the far-off thunder of the surf. We stopped in our tracks.

"You hear that? A plane! God help the poor beggar who's flying in this soup!" Payload turned to me, "You hear it too?"

"Yes. There it is again. It's coming closer," I had to listen to figure out the plane's direction.

"Let's go back to the dock. It's a plane all right. If that guy lands here all right, he's an artist."

The fog was thicker than ever. Like a curtain it hung over the Snye. Louder and louder grew the roar, suddenly it came from right overhead hammering into our ears, receded again, cut out, and then there was just a great rushing sound and a big shadow glided onto the ice. A burst of motor noise, and through the foggy curtain appeared the plane, another Canadian Airways Junkers.

"Must be Con!" remarked Payload. "That was some flying!"

The plane was tied up. Con crawled out of the pilot's compartment. "Get that fellow out! Get him to the Doctor!" is all he flung at us bystanders, walked off a way and began to pace up and down furiously, all the while getting rid of what he had on his chest. Strong language is no rare thing in the North, but all that had ever been said in unmistakable terms fade into a pink tea conversation, measured by Con's delivery. All present stayed at a respectful distance, somewhat perplexed, then bethought themselves and offered a helping hand to get the stretcher with a very sick looking bit of humanity out of the plane, and over to the Canadian Airways shack, where the young medico from Fort McMurray had just arrived.

Meanwhile Con had relieved himself of all he had to say about people who can't make up their minds, and then we got the story. Shortly after we had pulled out of Goldfields, it had been decided once more to fly the sick man out. All the planes were gone. What then? By wireless Fort Chipewyan was contacted, and Con, having landed there just for refueling on his way south had to make a detour of some 120 miles to make this daring mercy flight, the last leg of which must have been hellish. No wonder he was good and mad.

The poor patient, whose strangulated hernia was the cause of all this looked as if he had given up hope of getting through alive. Gently the stretcher was put on a toboggan and slowly and carefully we made for the Doctor's home, where the patient would spend the night, competently taken care of.

No long bull session at the Franklin Hotel. After a good supper, drowsiness overcame us all, it had been a long day, and so to bed. Only around noon the following day the fog finally thinned out sufficiently to permit a take-off. Con lost no time getting his patient to Edmonton.

We had to delay our departure. Word had been received that there would be two more passengers for Edmonton. We waited and waited, and then it was too late. Fog, thicker than ever, dimmed out the meagre daylight.

So, we sat and read and talked and had the odd beer. Our new passengers, a very charming young couple kept us company. It was their first trip that far north, and only a short one. They had attended the funeral of a brother, with whom we had had a few small beers when for the first time we had arrived at the Franklin Hotel. How long ago was that? Not yet three years! The most eventful time In my life. The North had given it to me. I was sure I wanted more of it; this could be only an unavoidable interlude.

We had to go back.

Shortly before we finally took off for Edmonton the following forenoon, a message came from the RCMP post that a sick woman had to be taken on, some little way up the Athabasca River. The spot for landing would be marked with spruce trees stuck in the snow on the river.

That's what Mac told me. Sounded very simple. All in a day's work. How he found the spot I didn't know. I tried to find the spot through the cabin window but the whole country was full of Christmas trees, sometimes one couldn't even clearly tell where the river was under that immense white blanket dotted with spruce, birch, and poplar, all just black silhouettes from high above.

But down we came safely, and right alongside a small group of people, among them a Mountie, not in a red coat though. His breeches told us. He must have dropped in at the trapper's shack on his patrol, and somehow got word to Fort McMurray which told of the woman's plight in this forlorn and chilly place.

There was no time for much talk that morning. Quickly, among the freight, luggage, and bedrolls a fairly comfortable place was arranged, and with everybody lending a hand, the pale looking woman and two somewhat bewildered kids were brought aboard.

Once more with thundering motor we rose into the sky, now definitely Edmonton bound. Finally, through the ground haze one dimly saw the upper part of the MacDonald Hotel, and the Tegler Building, a wide sweep into the wind, down we went, barely feeling when the skis touched the snow. The long flight was over, and once more, Lotti and I were together. Thoughtful friend Wop had brought Lotti right to the airport!

We didn't let the uncertainty of the future burden us too much. We were happy to be together, and as we were fortunate enough to find an apartment in the Highlands, there was nothing to prevent us from enjoying life by ourselves or in the company of good friends. Every day and hour had to count, because we knew that with the coming spring, another separation, a harder, longer one, was likely. Lotti was pregnant and being separated would be difficult.

Looking ahead then, one big question loomed up: Would the baby arrive before I had to go north again? Dr. Terwilleger said around the first week of May. If we were to go prospecting again, and make the season as long as possible, then I would have to leave early in April, before break-up set in at Edmonton.

It would come much later at Yellowknife, which, we had no doubt, would be our next field of operation, but if we waited that long it might be June before we were cracking rock again. Many ifs cropped up when we thought of the future, too many in fact to worry about all of them, but one of them couldn't be overlooked. Darker and more menacing were the clouds that loomed on the political horizon, the stock market was in constant jitters, definitely boding no good as far as our prospecting plans were concerned.

Deep into the night we often sat with Maud and Pat, trying to interpret the signs of the times, without getting much nearer to the truth, and often the best answer seemed to be: "To hell with all that. Let's have a drink and be merry!"

It wasn't that we were not inclined to face the issues, or that we didn't find any solutions which looked sensible to us. Looking back, I feel sure that we were more often right, than most of the so-called statesmen and politicians put together. We looked at things in an uncomplicated manner. We knew that we never could have had peace in our camps if we had let somebody play the bully, without challenging him, and therefore had little understanding of anything that reeked of appeasement. Uppermost in our minds though, and so discussed again and again, was the North, the endless stretches of undefended virgin country where we, ourselves, had made a few tiny scratches and found unmeasured wealth below.

Why wasn't something more done to really make The North a part of Canada, to make the exploration of it the supreme challenge, to conquer it and guard it before somebody else would get the idea of breaking into the treasure-house? Would only another war awaken Canada? But it was the beginning of March 1938, and still, we didn't know. Would our old outfit or another one send us prospecting? By and by we grew anxious to know, and this feeling increased with the decrease of our financial resources.

Many a time Pat and I met downtown to map plans. Finally, word came that some of the backers of our former prospecting activities had formed a new outfit: The Constellation Mining and Exploration Company, and two parties of two men each were to go to Yellowknife, before break-up, and later Pat would follow with a new plane yet to be acquired.

This was welcome news, but it was not easy to tell Lotti that soon we would have to part, about a month before the baby's arrival. How could I possibly go and leave her to face it all alone? I don't know what answer to our problem we would have found, by ourselves, but suddenly and miraculously the picture was brightened, the gloom dissipated, by good friend Fran appearing on the scene like a heaven-sent angel. "Mother and I want Lotti to stay with us till the baby arrives, and later, well, we shall see. The Doctor is all for it too, so there."

In no better or more loving, selfless care could I have left Lotti, and with a somewhat lighter heart could I think of the parting coming now so close. The short, busy days, filled with preparations for the trip into the bush flew so rapidly. Two canoes, tents, tools, etc. were purchased in Edmonton and sent by rail to Fort McMurray. Food, sufficient for the break-up we were going to buy from the Hudson's Bay Post at Fort Smith, from where our two parties would be flown separately into the Yellowknife district.

One morning Pat and I met downtown to do some shopping, which, of course, included a few cool beers. The peaceful occupation of consuming the refreshing stuff was rudely interrupted by an excited fellow barging into the beer parlor, "The Nazis are marching into Austria!"

Somebody turned on the radio. There it was. Pat turned to me, "Maybe you won't get to Yellowknife! Holy cats, this means war!"

"I don't think so, not yet!"

"Don't be stupid, man! The world will never stand for this!" Pat hollered.

"I am afraid the Germans will get away with it. Quite a few of the Austrians will welcome them, however sorry they will be in the future."

"Ah, nuts to that. They have to be stopped. You'll see. Let's go and see what the stock market says!"

There was plenty of excitement at the stockbrokers. War, war, war was on everybody's lips.

"Hear that? What did I tell you? Bet you ten bucks there will be war!" Pat was sure....

"I'll take it. Sorry to take the money away from you though. You feel there should be war, I understand that. But it won't be yet!"

"Just because you don't want it to happen? Holy cats, man, you'll have to face the facts."

"Pat, I am doing just that. My feelings are about the same as yours, but they don't count much in power politics. So, I'll go home now, but I'll have my dunnage bag packed and ready for the North."

The storm didn't break. Austria was no more, having been invaded and incorporated into Germany from March 11th to 13th, 1938. And, the world believed the assurances of no further territorial claims. A big mistake.

Late one night the telephone rang. Pat was at the other end, "Congratulations! You should be in politics. The war is off for the time being. Still want to go north?"

"Didn't expect me to change my mind, did you? If we don't get out soon, we'll have to go on relief."

"OK. Meet you at Canadian Airways tomorrow morning."

"We'll fly you in as soon as the weather clears," said Wop, and in an aside, "I'll get the news to you somehow, Hermann. I'll keep in touch with Doc. T." That was Wop again, never failing to think of the other guy, a great friend to have.

Prospective mother and prospective father then paid a last visit together to kind, fatherly Dr. Terwilleger.

"Everything is going fine. There is no reason whatever to worry. Babies are born every day, and you," he turned to me, "this may trouble your ego, but actually, you aren't very important, now. I am taking over. Now run along you two and enjoy the days you can still spend together."

CHAPTER XXX
DEEPER INTO THE NORTH

A FEW MORNINGS later someone called and the telephone shrilled, "Canadian Airways. Your plane is going out this morning. Come on down right away."

A last embrace, a last goodbye, parting was an extremely bitter sorrow. Ronny and Woods showed up soon after my arrival at the Canadian Airways Ltd. Office. Those two were going to make up one team. Art, who had returned to Moore Lake to keep Dan company was to join us at Fort Smith. He would be my partner, and I was glad about that. I knew we would get along together.

Packsacks and eiderdowns went into the trailer, we climbed into the truck. After crossing the Low-Level Bridge, we soon left Edmonton behind, heading for Cooking Lake air base. A strong wind was blowing, tail end of a late blizzard, piling the snow in drifts. Not a friendly morning. The skies were low and grey over Cooking Lake.

Quite a number of planes were lined up, mechanics checking motors, refueling, loading, and unloading. Last rush before break-up. Our Norseman just taxied in a wide circle to smooth some of the drifts and to polish the skis, then we loaded our belongings and crawled aboard.

"All set?"

"OK," the same conversation between mechanics and pilots was hear all down the line.

Harry gave her the gun, two fellows rocked the Norseman by the wing struts, then ducked to escape the clouds of snow the slipstream blew up. Into the wind. Faster and faster, we glided over the wavelike drifts. It was kind of bumpy. The motor sang a higher note as Harry pushed the throttle forward. We were up. Rocking slightly from side to side we climbed, and so we were heading once more into the land of the pale blue snow.

For the first half hour the weather wasn't any too good. Occasionally we flew through snow and often the ground was hidden by fog. When we saw far below us the valley of the

Athabasca River, the grey overcast was breaking up, patches of blue sky appeared, sunshine played over the silvery fabric of the Norseman's wings. For a while, our plane rocked around in the bumpy atmosphere, until clear blue skies were smiling all around. On an even keel we soon reached Fort McMurray.

From here a brave old Junkers, piloted by Don would carry us farther north. Hurriedly we transferred our belongings from the Norseman to the Junkers, in whose roomy fuselage there was also plenty of room for the other supplies sent ahead by train. Our two new canoes were securely tied underneath the wings, the gunwales fitting the wings' contours exactly, as if made so on purpose. There was just enough time left for a hasty luncheon at the Franklin Hotel.

Some friends were around when we were ready for the take-off. A last handshake, "Good luck, old boy."

"Thanks, Con. Same to you. Hope to see you sometime."

Again, we were on our way, Ronny and I sprawled on top of our supplies. No seats in the Junkers; every pound counts. Mighty Athabasca now a wide, white meandering ribbon threading its way through the black bluish and white crazy quilt below, was our guide. Northward, ever northward we journey, the motor singing its deepthroated singsong and the all-metal fuselage humming the accompaniment while the ends of the big, tapered wings are just tipping above, then below the hazy outline of the far horizon.

Gradually the pattern of the scenery changes, the country was flatter, less wooded, and though we couldn't see it under the white blanket, we knew we were now above the Delta of the Athabasca, where the big stream splits into innumerable small watercourses meandering through the mudflats into Lake Athabasca, whose immense blinding white expanse we glimpsed to the east. Not a sign of life betrayed itself down below, but there were times when the delta was a livelier place. When Indigenous and White trappers went after the muskrats, and in the spring and autumn when countless migrating birds rested there unmolested on their flights to and from the northern breeding grounds.

Leaving Fort Chipewyan to our right we picked up the Slave River, the left bank of which here forms the eastern boundary of the Wood Buffalo National Park, where the last remnants of the once mighty herds of buffalo found peace to live and to increase. Almost two hours flying time had elapsed when we passed Fort Fitzgerald. A little farther west and north the buildings of Fort Smith came in sight. Between the two points, where the waters of the Slave cascaded over a series of rapids, we noticed the sixteen-mile-long road of Smith Portage along which north-going freight was hauled by trucks or tractor trains.

As we came in for the landing at the airport, an enormous flock of perfectly camouflaged Ptarmigans rose into the air, and our pilot had to circle once more to get away from them and bring us safely down, near a few dilapidated shacks and piles of gas and oil drums. A pale-yellow sun dropped low in the west, as we made our way to the Hudson's Bay Company's Hotel Mackenzie, where the night was to be spent. Art, who arrived a few days previous came to meet us. Supper in the big dining room, a drink, smoke, and a little talk and off to bed.

The next night we'd sleep in our tents somewhere in the bush. Bright sunshine awakened us the next morning. At the Hudson's Bay Store, we gathered our food supplies and had them carted to the plane. Art and I were going in on the first trip. Don, our pilot showed up, "Where, exactly, do you fellows want to go?" I brought out my map to point out the approximate neighborhood Art and I have chosen for our first camp.

"Take anybody in there this season?"

"No, not yet."

"That's fine, and listen Don, when we get there do a little circling. We'd like a good spot for camping. You know, level ground, some fair-sized timber, some dry wood, some birches, and a low shoreline."

"Quite a large order!" Don grinned. "All right, we'll try. Hop in, one of you can sit up front with me."

The mechanic cranked the starter, the nose turned into the wind, and we were off, following again the Slave River until it took a swing to the west, but we kept to a course straight north to hit the settlement, near the south shore of Great Slave Lake. A short stop for fuel in a windy, desolate-looking spot. Some cabins, built from squared logs plastered with mud, a Hudson's Bay building, and a Mounted Police Post, that's all.

The Mountie and some Indigenous men came running towards our plane. They must have been glad to see somebody to the have the loneliness broken, if only for the few minutes it took to fill our tanks. It was my turn to sit with Don, and I was glad. I was chilled to the bone. Spring seemed awfully far away.

Across the eastern end of Great Slave Lake, where large reeflike islands and peninsulas stretching roughly east and west broke up the big body of water lay our course on the last leg of our journey.

Don cast an occasional glance at map and compass. We crossed the last long island, the last channel of the big Lake, then, "This is it!" He had hit our chosen lake dead on. "How about circling near the outlet, we should find a good spot there."

Flying in a wide circle we looked over the whole shoreline of the fair-sized lake, but near the south end we spotted a piece of level ground, the end of a ravine cutting through a rocky range. There was some green timber, fir, jackpine and birch in the ravine, and on the hillsides charred trunks of the burnt over bush stand black against the glittering snow. It was just what we were looking for. Another lower turn, the shadow of the Junkers grew big over the ice. We had arrived.

CHAPTER XXXI
WAITING FOR SPRING

QUICKLY WE UNLOADED and started clearing a space for our tent. There was no time to lose. By nightfall, the camp must be up. Out came our axes to bite into the frozen trunks of fir and pine, bringing them crashing down, branches breaking like glass. "What size tent we got?" called Art.

"10 x 12 with a three-foot wall."

"Three rounds of logs will do it then. That will give us plenty of room."

A dozen logs were soon cut to size, notched, and spiked together to form a solid frame. Now for some poles, ridgepole and tie poles and up went the tent. The bottom end of the walls was then tacked to the upper log, and while Art piled snow against the walls, I put up our tin stove. A quick dry wood fire soon melted the snow in our coffee kettle and into the pan went the canned sausages. A loaf of bread, smuggled into my bedroll by foresighted Lotti came in handy. Our appetites left nothing to be desired.

"Some more java, Art?"

"Yes. Thanks! Have a smoke. These are the last tailor-mades! From now on it will be makings again. Good old Ogden." Darkness had settled over the earth, but in the skies the brilliant fireworks of a clear northern night were shining. Stars sparkled like diamonds, and to the south the aurora borealis was weaving gossamer curtains. In our tent a few candles spread a friendly light, a fire crackled merrily in the tin stove. We spread the tent fly on the ground, blew up our air mattresses, and unrolled our sleeping bags.

"We just made it!" said Art.

"Yea, just about. By tomorrow night it should look a lot better though, after we build us some bunks and a table. Got to get the fly up too."

"How about a floor from poles?"

"Good idea, Art! We'll have a swell little place here, well sheltered and all."

So, we crawled into our sleeping bags. Lights out. For a little while we talked. Said Art, "Well, we are in the damn cold waiting for another break-up. How the years have passed! This is our fourth summer of prospecting."

"Looking back, it doesn't seem that long," I replied. "Maybe it's because our days were so well filled out. Always on the go, lots of work, plenty of fun and hardships too. You know, Art, I have often thought of writing down what's left in my memory of all those events that made up our life in the northern bush. Time will be long while we are waiting for the ice and snow to go. Later on, there will be rainy days to while away. I'll begin tomorrow." And that was where and how the first pages were written, by candlelight in a tent, at the south-end of Desperation Lake, east of Yellowknife, N.W.T.

Somewhat chilled and stiff of limb we crawled out of our cocoons in the morning, twiddled ourselves a smoke in a hurry and got our fire going. A big kettle of coffee was set to brewing, a pot of mush next to it. Some hefty slices of canned bacon went into the pan, followed by a couple of fresh eggs. Our small supply of henfruit wouldn't last long, but this was the first breakfast in the bush. The last of the bread gave us a few slices of toast. That kept us going for a while.

"Tell you what," said Art, "if you will do the cooking, I'll do the chores, get water, cut the wood, wash the dishes. Cripes! I'll do anything as long as you prepare the grub." That's how I became the cook of our team, but the dishes we washed together. I never heard any complaints. After four or five cups of coffee (we both liked our java) Art went to work cutting small poles for the floor, while I got busy fixing up the inside of our mansion.

When we were through, this was what our quarters looked like: Lifting the tent flap, to the left sat our tin stove, to the right an arrangement of opened packing cases neatly piled up, holding food supplies. In front of that a narrow table made from the packing case covers. Then to the right and left on a level with the top log, our bunks, made from a pole frame with burlap bags nailed to them. Between the two bunks, at the head end, a table made from packing case boards. Above that against the tent wall hanging shelves to hold books, ammunition, our cameras, and tobacco.

It was a neat and compact arrangement, and the time spent on creating it was well repaid in comfort. In our later moves, we always stuck to the same scheme. Once having made our tables and shelves, it usually took us only two hours to repeat the set-up when we settled in a new location. It earned us the title of "Dudes" bestowed upon us by other fellows who happened to drop in at our camps.

"Might as well be comfortable," was our standard reply. We knew we had a good thing. The pole floor was a great arrangement, always clean and dry. Later on, we even had a bathtub.

The balance of our food supplies was safely stored in a cache, a triangular platform high above the ground between three stout trees. "What's in that other packing case, and in that bundle over there, Hermann? More grub?"

"I don't think so, must be the hardware, pails, etc. Let's check it."

We sure got our surprise when we opened the box. "What in hell is this for? Well, I'll be a son of a gun, if it ain't a pair of horse clippers!" Art was waving a long-handled pair of clippers. "What do you know, prospecting on horseback! Them city slickers must have fancy ideas. That's a good one. Horse clippers, ha, ha!" Art was rambling on. Then it occurred to me.

"You know what happened? I remember now. When we ordered the stuff, we asked for two pairs of hair clippers, for humans, that is. Tough guys! Well, maybe they think we are real tough guys!"

"Two pairs, you said? Well, here you are, brother. Two pairs of genuine horse clippers, and not a bloody horse in a thousand square miles!"

Fortunately, the rest of the stuff in the box was closer to what we had ordered, but it was plain to see that whoever had filled the order had never spent a day in the bush. If we needed any further proof, it was revealed when we opened the remaining long bundle. After the sacking was off, we first saw a big sheet of galvanized tin rolled around the bundles of tools. "Woods must have ordered that!" said Art. "Last year at Gordon Lake he was always talking about tin for sleigh runners! I'll be a monkey's uncle if I ever build a sleigh again, but that's a long story. Tell you some other time!"

"Yea, I'd like to have your end of the story of what happened last year. As to that tin, I have an idea it may make us a bathtub. We have to figure that one out."

We pushed the rolled-up tin off the bundle and looked at each other. There was something for an amateur gardener, and a bundle of mattock heads that wouldn't help much as our picks.

"Oh, well," said Art, "let's quit prospecting right now and go farming. We got all the tools. My old man always wanted me to be a farmer."

Fortunately, we both had brought a prospectors' pick, which would last us till after break-up, so we could laugh about the strange collection of tools we had unpacked. The day was almost over when we had everything shipshape. Fried dried spuds and Mulligan stew made from corned beef, canned tomatoes, dried onions, and a hefty dose of Keen's Mustard tasted good after a long busy day in the crisp winter air, so did the stewed dried apples, peaches and prunes, and the big potful of strong coffee.

Art was soon lost in a Western novel while I, two days late, unwrapped Lotti's birthday present dug up from the bottom of my packsack, *Anthony Adverse* by Hervey Allen. A happy choice! I was deeply touched and for a few moments I had to fight to hold back my emotions, my yearning for the one I love that welled up from my heart. I forgot about the reading and took my pen to talk to my love from a full heart, if even only on paper. Art looked up. "What are you writing? Start our story?"

"No, Art. I am writing to Lotti."

"No use now, Hermann. Break-up is a hell of a long way off."

"Just the same, Art. I have to. Look what Lotti gave me for my birthday."

"Your birthday? When was that?"

"The day we left Edmonton. Seems ages ago."

Art looked at the big tome, "1300 pages. Cripes! I would never plow my way through that. Don't cotton much to them big books. Give me a Western adventure, that's what I likes."

"You will like this one, I bet you, it's said to be full of hair-raising stuff."

Off and on the fire crackled, a candle sputtered, that and the scratching of my pen made the silence the more pronounced, until suddenly there was another sound, deep and rumbling, startling me. Art had dozed off, snoring lustily.

"Hey, there, Art. Better get into your eiderdown."

CHAPTER XXXII
AT THE OLD GAME AGAIN

COLD, BUT CALM and clear weather was still with us, and we made the best of it, exploring the rock along the shoreline on snowshoes. The country was under a thick blanket of snow, but walking on the ice, we could at least investigate the cliffs to get an idea of the geological formations. In the Yellowknife District gold had been found in quartz veins and outcrops which had penetrated the broken up, and often much altered old sediments. These we found here too, also numerous quartz veins but not many signs of mineralization.

However, just going along the shoreline could not be called thorough prospecting, but it might show us the direction in which to work once the snow had gone. We had been at it for about a week, when one morning a sound which we hadn't expected to hear again for at least another five to six weeks called us out of our tent the sound of an airplane flying low and nearby. It was a great surprise and pleasure to see Harry land his AXE, an old familiar Bellanca, down near our camp, and it was good news to hear from Harry that he had been posted at Yellowknife for the break-up period.

Hastily we scribbled a few notes for the folks at home. Even at Edmonton winter hadn't loosened its hold yet, planes were still flying on skis. While we were still talking to Harry, a sudden change had come into the atmosphere; clouds, grey and heavy rolled in, blotting out the sun, the cold seemed more penetrating.

"I'd better get out of here. Looks like a blizzard. I'll try to get back here once more before the ice goes. Take care of yourselves."

"Thanks for calling, Harry. So long!" we called out.

Towards the north end of the lake he raced, soon swallowed by the grey bank of snow clouds which came rolling towards us. In a few minutes all was blotted out by a full-sized howling snowstorm. Winter had won another round. When the skies cleared, the cold was more severe than before. 16° below said our thermometer on April 22nd, but soon after that the change came rapidly. Shifting winds changed to steady, balmy breezes from the south.

Daylight came at 6 a.m. and old Sol travelled in a giant circle almost back to its starting point where it arrived by 8 p.m. The snow began to melt during the middle of the day, but in the mornings, there would always be a crust strong enough to support us on snowshoes. Flocks of ptarmigans appeared, but never came close enough to end up as a roasted treat. On May 1st, Harry showed up for the last time on skis, making dark, wet tracks on the collapsing snow.

This time we had our mail ready, and a list of things we would need after break-up. Harry said he had seen caribou going north not far from our lake. We were hungry for fresh meat, but three or four days passed before we saw the first herd cross the lake a couple of miles to the north of our camp, just grey shadows in the distance. Art cleaned and oiled his 303, and I gave my Savage 22 a going over. We hoped to use them soon.

The outlet of Desperation Lake where our camp was situated, was a long narrow channel which, at the north end very abruptly widened out into the main lake. One morning, we decided to explore the south shore. Rounding a small rocky peninsula, just before reaching the main lake, we suddenly saw a grey and black indistinct mass lying on the ice.

"What in hell is that?" we both exclaimed.

"Dead caribou? Maybe?"

"Look, it's moving." And so it was, but only the black part.

"A bear. Let's get closer." I got out my Reflex Camera, while Art held his rifle at the ready. The wind was against us. Bending deep down, cautiously, step by step we moved closer. Just about 100 feet from the tugging and tearing bear I took the first snapshot. Maybe the sound of the shutter made him rear up; he was a big fellow. What a picture! Hastily I turned the film, too late. Art's gun barked, but in a few mighty heaves Master Bruin reached the shore.

Again, Art fired, then our intended victim was gone, but we heard him plowing through the dense bush, climbing the steeply rising ground. "Cripes, I must have hit him," cried Art. "Let's go after him!" There was no blood to confirm Art's hopes, and the speed with which our bear travelled surely was no indication that he was hit.

Did you ever try to catch up with a bear running for dear life up a steep hill? I only say, "Don't!" We tried it, till our tongues were hanging out, and our chests heaving. Only one more glimpse we had of him, mighty hind legs pumping tirelessly. We had to give up, we were steaming.

"Damn funny though, I didn't hit him with my first shot. I sure had him in my sights!" We should have investigated further right away.

"Are you hot?"

"Am I? Brother! By the time we get a chance for a dip we'll be good and high!"

"What say we go home, and I make us a bathtub out of that piece of tin?" I suggested.

"Sounds good. How in hell you'll get a bathtub out of that I can't even guess."

So, the mighty hunters trudged home, and I went to work. It went better than even I expected. Turning up all four sides to form right angles with the bottom, using the back of my axe and a prospector's pick was no great job. To fold the corners was a bit tricky, but with Art giving a hand, we managed. Now we had a somewhat wobbly pan about 2 ft. X 3-1/2 ft. and approximately 12 inches deep, but the rim was pretty sharp. We had no wire, but a length of clothes-line cord offered a solution. A pair of blasting cap crimpers had to serve as pliers. With patience and perseverance, I neatly folded the rope into the upstanding edge of the tin.

To call the product a bathtub would be saying too much, we called it a sitz-bath, but it surely served us well, and Art paid me the greatest compliment. Long before I was through making the safety edge, he fired up to heat a few pails of water. Pretty soon the two of us smelled of Lifebuoy soap, which was supposed to be socially acceptable.

"Honest, Hermann, I was beginning to worry that somebody might whisper dreadful things behind our backs, things your best friend won't tell you, or something. But then again, who is there in this blasted wilderness to whisper. No fooling though, it feels better. Boy, I am hungry. What's on the menu tonight?"

"Well, we almost had bear steak. How about pancakes for an opening? After that I will try to bake us a cake."

"No kidding? What kind of cake? I could go for a hunk of chocolate cake, with icing an inch thick!"

"You will get your cake, but you'll have to run to the corner store for some icing sugar," I quipped.

Every second day I had baked two big pansful of some sort of glorified bannock. The recipe was simple: Equal parts of white and whole wheat flour, some cornmeal and oatmeal, a dash of salt, some sugar, dried eggs, powdered milk, a liberal dose of baking powder, all this sifted together dry, enough water to make a dough, some melted fat, and plenty of raisins. Real cooks may shake their heads about this, but we liked our bannock. It always rose nicely, had a nice brown crust, and best of all it kept fresh and moist.

On all our exploration trips we packed big hunks of it for our noon-day meal, and often it would finish off our supper with some jam or honey. The chocolate cake was developed in about the same manner, only the corn and oatmeal were left out, and part of the volume made up by cocoa and sugar. Tasting the dough, I tried to remember what it tasted like when, as a kid, I scraped out Mother's mixing bowl. It seemed all right. Into the oven it went. Soon enticing smells began to fill our tent, escaping through the not very tight door of the baking oven.

"Must be done by *now*." Every few minutes Art sniffed, and uttered fears that it might get burned. It looked fine when we finally took it out and smelled delicious. I must have been a little bit heavy on the baking powder though, there was a little too much ventilation in the texture.

"Tastes good just the same," Art remarked.

"I am glad you took over the cooking. Didn't count on this at all. Let's make another pot of java to go with it. Nothing like living in style. May I pour you another cup, sir? and all that sort of thing!"

Some days later, it must have been on a Sunday, because in true bush manner we had been washing socks, shirts, and union suits in the morning, then treated ourselves to a bath and clean togs. Art got restless after lunch, grabbed his rifle, and made off, promising fresh meat for our supper, "I'll get us a caribou if I have to follow it to the Arctic Circle." Off he went.

After a long, long while, I thought I heard a shot way towards the North end of the lake. To do my share, I had put another chocolate cake into the oven. When it was done, I stepped outside. There, hardly fifty feet from the shore five caribou were standing, a perfect target, and poor Art was gone! Not a sign of him yet. Maybe an hour later, a dark spot on the horizon slowly grew and came nearer. Seemed oddly shaped, sort of top heavy.

At last, I could make it out. It was Art with the two hindquarters of a caribou still connected, draped around his neck. He was jogging along, with his head pulled down by his gory neckpiece, his shirt, and breeches, in the morning so nice and clean, now soaked with blood and sweat. Puffing and heaving he threw down his burden near our tent. He still hadn't seen what was going on in our neighborhood.

"I hate to do this, Art, but just turn around and have a look!"

"I'll be a son of a gun! I travels all over the bloody North country to shoot me a bloody caribou, and here they are just ready to be knocked over right at the front door. I'll be damned! One never knows, does one? Now, I'll even have to have another bath."

"I'll have the steaks ready when you are through with that, and cheer up, there is a chocolate cake waiting too." The fresh meat, so long missed, was a great treat, the knowledge that it had been hard gained added spice, and life was good.

CHAPTER XXXIII
ON THE GO AGAIN

DAY AFTER DAY the weather became more springlike, the rapidly disappearing snow permitted closer prospecting, and we were constantly on the go, using the ice as short-cuts on our trips. Going inland, from the south-east corner of the lake we found the country much broken up and crisscrossed by an abundance of quartz veins. To our disappointment, however, these proved to be singularly free of minerals, at least as far as could be determined by trying to break into them with our prospectors' picks. Dynamite was called for, but for that we would have to wait till after break-up.

Systematically we worked our way around the lake, and so one day we got to the north end from where higher hills had been beckoning us. They turned out to be barren granite, nothing to excite a prospector's mind, but along the foot of the range ran a sand-filled ravine with a small stream meandering through it. Trees had grown there to quite an unusual size, and with the sunshine filtering through them, it wasn't hard to picture it as a very romantic spot once the grass, and the foliage on the birches and poplars would be green again.

We had reached the draw some distance inland. When we decided to return home, as the sun was sinking lower, we followed the stream towards the lake, where the draw grew into a sandy plateau. We weren't the first ones who had found the place attractive. Scattered all over the place were neat geometrical arrangements, weather bleached, of teepee poles, just as the Indigenous men had left them, who knows how long ago.

We forgot the rocks for a while and poked around in the leftovers of old campfires. There wasn't much to give us any clues about the travelers of the past, a few charred sticks, pieces of bone, and half buried in the sand, an old rusty cast iron kettle.

"Must have cost the man plenty at the Hudson's Bay Post, Art!"

"Yea, bet you the person who forgot it here caught hell!"

It was our last long trip for quite a spell. On our way home we ran into a sudden rainstorm, soaking us to the skin before we reached camp. It was the beginning of a rainy period, shower after shower drummed on our tent fly. Inside we were snug and warm. Our pole floor now proved its worth. The ground was soggy, a million little streams making their gurgling way to the lake. Off and on the sun would break through the low-hanging clouds producing magnificent rainbows. Would we find the pot of gold at the end of one of them? There were so many shifting with the wandering sun, each one more brilliant than the last one.

So much ground to cover, somewhere there would be a hidden bonanza. To find it, a rainbow would be about as good a clue as the "prospectors' hunches" we usually followed. While the drenching rains streaked down from the leaden skies, holding us in camp, the enforced idleness nourished the worrisome and longing thoughts that had been with me every day since we went into the bush: How was Lotti faring? Had the baby arrived? How was my brave little woman now? I summoned all reassuring facts to chase the gloomy thoughts that kept creeping in; I recalled good Dr. Terwilleger's words, motherly charming Mrs. B. and brisk, competent Frances, a marvelous trio of friends who were going to take care of my loved one, and thus I found peace and hope for a good outcome, but the yearning remained.

It was easiest to bear when we were on the go. Sitting in the tent, waiting for the rain to stop was tough. So, I picked up *Anthony Adverse*, while Art was enjoying his Westerns.

Rain, rain, rain. Water was standing deep on the ice, still anchored to the shores. Great herds of caribou were crossing the lake splashing through the water. Their timetable must have been out slightly, it wasn't the best time to travel any more. Slipping and sliding, they often bumped into each other. When their feet went out from under them, they scrambled furiously to regain their balance. But on and on they went following their inborn wanderlust.

Long before I had come to the last of *Anthony Adverse*'s adventures, Art was through with his books and he grew quite impatient reading a few pages whenever I laid the big tome down, to look after our bodily welfare, to roast or fry some more hunks of caribou meat and prepare the fixings. Finally, he had it all to himself, and, I am sure, more eagerly a book was seldom read.

Art was lost to this world, rapidly the pages were turned. One day, about halfway through it though, I noticed that the rustling of the pages had stopped. Art seemed to be lost in thought.

"Where are you now with *Anthony*, Art?"

"He's having himself a time in Africa. Boy, oh, boy. Why didn't we go to some warmer place, where a guy can get himself a dame like that!"

"Neleta? Must have been quite a girl!"

It was high time for the rain to stop, for us to get going again, to wear off the surplus energy. The sun did reappear at last, and the country was stunning. In our rubber knee boots we trudged through the water that stood on top of the ice, going to the east shore once more to look at some quartz veins. When we passed the spot where we first encountered the bear the remainder of the caribou carcass was gone. It had been dragged to the shore, and not much was left of it.

"Son of a gun. He's still around! What say we get us another caribou, we are out of meat anyway, and use the leftover as bait?" The big herds had passed, but quite a few stragglers were still passing our camp.

Art went for his rifle, and soon we had what we wanted. The hindquarters, liver and tongue were for us, the remainder the bear could have, if he wanted to come for it. Halfway between our homestead and the opposite shore we deposited it.

Back in camp we were startled by the sound of an approaching airplane. My first thought was, of course, that it might be a Canadian Airways plane. Hadn't Wop promised to try his best to get the news to me if at all possible? The plane landed about a mile to the north. In record time we reached it, but there wasn't a soul to be seen. Then Art said, "If I'm not mistaken, that crate belongs to the McLaren brothers. They must be after something."

We looked around, there were tracks leading westward. After hanging around for a while, we decided it would be just as wise to go back to camp. Whatever they were looking for could easily be detected by us the following day. The time they stayed would give us a clue as to how far inland they went, and their tracks would be clear enough. Never overlook a bet, is a good motto for a prospector. We had been followed often enough ourselves. The party didn't stay very long, we were back at our tent only a short while when we heard the plane starting up again.

Early the next morning we travelled first west from our camp onto the higher ground, then straight north. Soon we ran into clear tracks. Whatever those fellows were after remained a mystery, unless they had been attracted by a series of large regular quartz veins, which probably had looked interesting from the air. We had been over the ground before without finding a trace of mineral, although it had been interesting in another respect; glacial action had planed the rock into a smooth plateau. The formations having an almost vertical dip, the different strata showed as clearly as a textbook picture.

To make sure we hadn't missed anything we travelled back and forth across the tracks and deeper inland, broke a lot of rock, but found no mineral, let alone the elusive gold. A busy day thus passed without time to worry too much, although now I had to give up hope for news before break-up. A few more days we could travel on the ice, after it finally broke from the shores, rumbling like thunder, around the middle of May. Then it floated on the swollen lake, moving slightly with the shifting winds.

It had become a habit with us, first thing on arising, to tie back the tent flap, to have a look at the sky, and cast a glance across the outlet. Right in line of view was the lure, the remains of the last shot caribou. One morning, I finally saw what we had been waiting for, Master Bruin was having himself a free meal.

"Hey, Art, get up. The bear!"

Pulling on his breeks, and grabbing the old carbine, was done in a wink. The morning's quietude was shattered by the rifle's roar. We thought we heard a dull thud when the slug found its target. The bear, aroused, reared up on his hindlegs. "Got him!" said Art.

But no, the bear quietly went back to his breakfast. Again, the rifle barked. Again, a dull thud, once more our intended victim reared up, then slowly retreated to the far shore. "Damned funny!" said Art. "I was sure I had him in my sights. Something cock-eyed about this here rifle. Thought so when I shot the last caribou."

"Let's investigate, it sure is no damned good this way."

Art pumped the lever to empty the magazine, pulled the trigger a few times, "Here you are, you tell me what's the matter with it."

The rear sight was all right, the elevation OK, but when I touched the sight, it seemed to move!

"Feel this, Art. I think we got your trouble."

"By golly! I was beginning to wonder what the matter was with me. How could we fix it?"

With hardly any tools and not the faintest idea how a craftsman adjusts the sights, we seemed to be up against it, until a saving idea dawned. I had a compass with a sighting arrangement. Bringing the rear sight in line with this, we determined the correct location of the front sight, or at least that was the theory we followed before we tightened the bead by carefully driving the metal of the slot against the base of it. A can lid then served as target and we tried our handiwork, Art could shoot straight again!

"Now let him come!"

He came all right, but apparently before we got up. A few mornings later the lure was gone. We investigated. It had been dragged behind a huge boulder on the far shore.

"Oh, well, he's the smarter one. He's got us beat," we admitted, and went back to prospecting.

South of the rapids which formed the outlet of Desperation Lake began a long, channel-like lake. Our map showed us that the trench-like body of water was several miles long. Near its end the Beaulieu River poured its waters into it. An ideal set-up to explore a lot of country by canoe, saving us walking many a weary mile. When we discovered that the ice had gone out of the long channel while it was still impossible to use our canoe on icebound Desperation Lake, we made plans to move south. Several trips would be necessary to make the move, even if we didn't take all our stuff. The remainder would be safe enough in the cache. We made our preparations, hoping for a south wind to drive the ice northward, thus freeing the ice-choked outlet. As soon as that happened, we would get going.

Days went by, the wind kept blowing from the north. When no change came during the day, we hoped to see it in the morning, so as usual, first thing on arising: lift the tent flaps and peek across the lake. Not the hoped-for open water but something else did I glimpse a few mornings later.

Coming straight for our tent the bear was leisurely walking across the ice. "Art! Quick, quick, get up. Your gun. The bear is coming!"

This story, as related here, happened much faster than I tell it. Art got up in a flash, didn't waste any time to tuck his shirttails in, reached for his gun. I took my 22, and my camera. Crouching low we snuck behind a clump of bushes near the shore.

"Let him come close, so I'll get some good pictures," I whispered to Art, who kept the bear in the sights of his gun. Off and on the bear stopped, lifted his head, ponderously turning it to sniff the wind. I took a picture. The bear resumed his walk, seemingly quite sure of where he was going. And so, he got to the edge of the ice. There were a few feet of open water, about twenty feet of rock-strewn shore, a clump of shrubs, and behind it are Art and me.

A few low grunts, a few sniffs. Suddenly there was a mighty leap, Art's rifle barked, I heard him hiss, "Dammit, I missed!" Out of midair the bear plummeted onto the rocks. There was no sound no struggle, just a silent black mass lying on the shore. We looked at each other, speechless at first. Then, "He can't be dead. The rifle just went off, I tell you!" said Art.

"Maybe he suffered a stroke," was all I could say. Evidently the bear was dead. With some caution we finally walked over. No doubt, life had departed, but how? There was no blood.

"Must have been heart failure."

"No, look here, Hermann!" said Art, "Nobody will ever believe it when we tell this story. Look!" Art had pulled up the bear's head, the shot has gone through both eyes. That bear never knew what hit him. To us, in a way, it was a sort of consolation, we had no grudge against bears so far. That came later.

"What'll we do now? Skin him? The fur looks good."

"Sure, let's. What's the best part of a bear?"

"Don't know. I've heard about bear ham."

For us inexperienced guys, it was quite a job getting the pelt off intact, with the animal lying on the rocks, but we managed somehow, stretched it between the poles of our cache, scraped the fat off the hide, and rubbed it thoroughly with laundry soap. While Art was still at it, I hopefully trimmed a roast off one of the hind quarters and pushed it into the oven. Supper was repeatedly postponed that night, but the longer I cooked the roast, the tougher it got. We ended up with caribou steak. Maybe that was just as well. There was an all-pervading smell of bear around us that night.

"Is it me, or you, or both of us, Art?"

"It's bear in any case. Couldn't smell any worse, where he was hibernating. We'd better tackle the lifebuoy soap again before hitting the hay, or it'll get into our bedrolls."

"If somebody told us a bear story, like the one we experienced today, we'd call him a damned liar, that's a sure thing. A bear jumping into a bullet. That's a real tall one," I chuckled.

"Yea. One never knows, does one? But you got some pictures, didn't you?" Art asked.

"Art, I have to confess. I have a strong suspicion that I didn't turn the film after each exposure. At least not always!"

"Excited, what? Well, you weren't the only one. That shot went off by itself, I tell you. Good thing it did though," Art said, shaking his head.

"Reminds me somewhat of our earliest days at Beaverlodge. Did I ever tell you? We had just built our first dock. I was down at the end of it one early morning getting some clear water. Suddenly the water boiled up as if a young volcano were at work below. I almost went into the drink, didn't know what was going on. I hollered, 'Pat, get here quick, bring the rifle!' I can still see him stepping out of the tent, whipping down his Lueger. A shot cracked. The commotion in the water died down, a reddish tinge appeared and to the surface floated the biggest jackfish you ever saw, with a half-swallowed smaller fish in his big jaws."

"Go on. Tell that to somebody else!" Art laughed.

"There you go. Who is going to believe our bear story?"

CHAPTER XXXIV

OPEN WATER

FOR MANY DAYS we had been waiting for a good strong wind from the south to clear the bay of ice so we could launch our canoe. What we got instead on May 27th came howling from the north, a full-grown snowstorm. It looked like deep winter again all around us. The fierce wind helped to break up the ice though but drove it in the wrong direction. The lake's outlet was choked with big floats, some were even driven, crashing, and groaning high onto the rocky shore. We had to wait some more, if we didn't want to pack all our stuff for at least a mile to the other end of the rapids. It was a good thing we didn't let our impatience run away with us, because winter's last stand was short-lived. From the south a strong, warm breeze came blowing, the snow vanished, so did the ice in the bay. Open water. Away we went!

We felt a tremendous relief. Winter was over. The paddles were flashing, and humming and singing, we pushed our heavily laden canoe to the south. Going down, we clung close to the western shore of the channel, watching for signs of mineralization, on our way back we would watch the east side. However, we didn't get far on that first trip. Soon we ran into ice, cached our load near the shore and returned to camp. North wind, south wind, they played a game with us.

"Just think in Edmonton the lilacs will be blooming, and here in this Godforsaken country, we still battle with the ice. Why didn't I stay home?" said Art.

"Yea! Just think, the baby must be three weeks old by now, and I don't even know how things are, whether it is a boy or girl!" said I.

"That's right! I shouldn't have brought it up. Brother, if I ever get married, I stay home. You won't see little Art traipsing off to the North!"

"Can't be long now, surely. Even to this country, spring has to come!"

So, we kept the relief valve popping, talking, grousing, and swearing.

It was in the evening of the last day of May when we finally got under way again, with our second load. By then we were deep in the time of the bright nights and long days. The dying and the newborn days were only separated by a short period of twilight. It was as if the North had slept enough during the long winter, and now just took a deep breath before every new day's dawn. The wind died down, and quieter than ever was the silent North. The lake was like a mirror in an unlit room. There was only the swish and then a tiny gurgle as we pulled our paddles through the lead-colored water, and once in a while the hollow sound of a paddle hitting the gunwale.

Close to the area we had figured would be our next stamping ground, we ran once more into solidly packed ice. We turned around and picked a small island we had just passed, to dump our load. A good safe place. We felt somewhat cold and tired, but a big pot full of strong coffee soon restored our spirits, and around midnight we started the return journey. Two hours later we arrived at the portage into Desperation Lake. With our canoe on high we trotted across. Nature had chosen this moment to put on a gala performance. The skies were on fire. Cirrus clouds and mares' tails high above our heads, tinted a rich golden red by the afterglow of the setting sun, and while we silently watched in wonder, the sun reappeared, sending yellow and orange streamers of light heavenward, to create a symphony of colour our eyes had never seen before. Slowly the colours blended, and the atmosphere seemed to fill with a soft, but almost tangible, violet radiance. Framed by the black and solemn silhouettes of the hillsides, the big lake was a gigantic reflector. Awed and silent two small figures stood near the water's edge; I don't know how long. Not a word was spoken, but I am sure we both felt it, "and heaven and earth shall proclaim His glory."

Suddenly the colours faded, the vision passed, and there was only grey dawn. A cool breeze ruffled the water. We shivered, jumped into our canoe, and covered the last few hundred yards to our tent and sleeping bags. Tired, but richer by an unforgettable experience, we soon sank into dreamless slumber.

CHAPTER XXXV
NEW CAMP

BREAKING CAMP THE next day didn't take long, but our canoe was almost down to the gunwales when all our gear was stowed aboard, and some stuff, not immediately needed had to go onto the cache. A few more tins, cut open, were nailed around the tree trunks to prevent bears from crawling up, and feeling quite assured, we shoved off. A couple of miles down the channel we were startled by the faint noise of a kicker.

"That's Ronny and Woods," says Art.

"They have no kicker, and it's a long way from Pensive Lake," I reply.

"Who else would it be? Don't tell me there are more guys than us crazy enough to stay in this neck of the woods over break-up! There, look! It's them."

Around the next bend, hugging the shore a canoe came in sight. Two fellows in it. We hailed them with lusty shouts, echoing from the steep rocky walls of the channel. At first there was no answer. They were almost up to us, when an elderly fellow sitting in the stern cut the motor, and let loose in a high-pitched British voice, "Oh, I say there, chaps! Where are you going?"

Art half turned around to me, spit out of the corner of his mouth, "What do you know, a bloody bloke!"

The two canoes glided along together. Art was close to the elderly fellow, who poured a long, involved story into Art's ears. I tried talking to his young, rather sullen looking sidekick, but couldn't get a rise out of him. The old boy was still pouring it into Art. When he noticed that I was trying to catch his words, he muffled his voice some more. Strange customer, all I overheard off and on, Ontario, Ontario. Where have I heard that before? At last, I heard Art, "Well, we got to get going again. See you some other time."

"Righto," replies the confidential type.

"What great secrets did that strange bird impart?" I asked Art when we have turned round the bend.

"Understood only half of what he was saying. He doesn't think much of this country. Wants to go back to Northern Ontario. Didn't want to tell for whom he's in the field. I got it out of him though, sounded like Oro Plata, or something. Ever hear of an outfit of that name?"

"Yea, has been in the papers. Did he meet any other parties?"

"He says there are quite a few near the Beaulieu River, where we wanted to go."

"If that's so, we'd better look for another good spot on our way down. After all, if we get lonesome, we can always go and see them, but for prospecting, we are better off by ourselves."

"That's right. There are always some guys who like to profit by somebody else's work. If they think you've found something, you can't shake them, and before you know it, the ground is staked solid around you."

Before long we had found a spot. That looked just right to us. The channel was somewhat wider here, looked more like a lake, and from the east shore, quite low, opened a wide vista into the country. It looked like a main break in the formation, and that was another good reason to stay a while and have a good look all around.

On a rocky point our tent was soon erected above the traditional few rounds of logs. No need to make a floor, the rock, old sediments, altered by quartz intrusions was perfectly level and smooth, planed down by glacial erosion. Out of a clear sky the sun was beating down. Summerlike heat was brooding over the land. Mosquitos were buzzing our ears, attracted by the freely pouring perspiration.

As soon as the tin stove was in its usual place, we had a good meal and a big pot of coffee, and by then the other unfailing camp guests, the whisky jacks, had found us too.

Art twiddled himself another shirt pocket full of cigarettes, twirling both ends tight, so the tobacco won't fall out. I lit my stinky old pipe, and then we set out to fetch the load of stuff we cached on the little island to the south. On the way back, we thought we heard an airplane way to the north, but we couldn't be sure.

"Maybe it's just Oro Plata coming south again," said Art.

"Well, it surely can't be long now. You know Wop promised."

We had been so busy; I hadn't had much time for worrying. But how I wished a plane would finally come and bring me the news!

We took our time getting back to camp; leisurely following the shoreline we cracked rock here and there. From where the sun was slowly sinking towards the northwestern horizon, once more we heard, and this time there was no mistake, the roar of an airplane drifting in waves. Sometimes it came closer, then diminished, and again it faded away completely, to leave us with just the sighing of the breeze and the war-chant of the mosquitos.

"Maybe they looked for our old camp! What do you say, Art, we go tomorrow and get our stuff from there? Maybe they landed and left a message."

"OK by me," answered always cooperative Art.

"The sooner you get your news, the better for both of us!"

"Does it go on your nerves too, Art? Or is it me? You certainly have been very patient. I want you to know, I appreciate that," I said.

"Aw, forget it. Hate to see a guy in a fix like you are in. I have been in a fix myself, but that was somewhat different. My old man is still paying for it."

"I don't have to tell you, Art, but I think your old man is a great fellow. I am glad I met him."

"Yea. I think the world of him, but I haven't always been an unmixed joy. However, he never forgets that he sowed some wild oats in his time. One thing I'm damned sure of, he would rather be here with us, than having to push to that old stuffy office, day after day!"

So, we talked for a while, as every evening before we crawled into our eiderdowns, and slept, blessed sleep took us into the land of dreams.

When we reached Desperation Lake, early the next forenoon, we found the whole outlet jammed tight with the last of the ice. There is no question, a plane couldn't have landed there. While we were still debating whether or not to pack our supplies from the cache overland, the world around us suddenly turned dark, towering thunderheads with sulfurous linings blotted out the sun, lightning flashed all around us, thunder crashed, violent gusts of wind from the north tore at the trees, and pushed more water into the outlet, where the piled-up ice heaved and groaned.

Sheltered by our overturned canoe, we watched the uproar. When it blew itself out, and the sun reappeared, painting rainbows into the sky with the help of the last showers, the bush became a steaming jungle. We hiked over to the old camp site but found no message. All else looked exactly as we had left it. The cache was in good order, and we decided to leave things as they were.

On our return journey to our new camp, we spent a few hours prospecting some interesting-looking draws that open eastward from the channel. Another day was almost over when our tent was reached. Art started chopping some firewood and I had started to get supper underway, when with dramatic suddenness a low-flying plane roared over our heads, heading northward. We shouted and waved, as if we could force him back! Ah, he turned at the end of the lake, and came in for a landing! My heart beat a staccato. At last, after two whole months in the bush, I would hear the news.

The water was too shallow near the camp for the plane. Art and I somehow tumbled into our canoe, and we paddled over to the Canadian Airways Bellanca for all we were worth.

The pilot waved to us, so did the co-pilot who had thrown open the cabin door and shouted, "It is a girl, Hermann, all is well!"

I must have said a silent prayer. We were up to the plane. Rudi Heuss was the pilot.

"Wop was after us all the time to get word to you as soon as we could. North tried before, too much ice on the other lake. Wop says it's a swell baby, and Lotti is fine! Happy now?"

"Yes, by God, I'm happy. Thank you, Rudi, and give my thanks to Wop," I enthused.

In the meantime, the co-pilot dug out our mail and several parcels. Art caught them and deposited them in the canoe. Two bottles of beer appear, happy surprise, were opened with an expert flourish and handed down to us.

I was so anxious to learn more details of the blessed event that I began tearing my letters open. Absentmindedly, I put the bottle on the gunwale. Don't ask me why, and don't ask me how it was possible that the bottle still stood on the gunwale when we went ashore after Rudi told us to get out of his road. He had more calls to make. I grabbed the bottle just in time as Art hoisted his, "Here is to the new daddy! How's it feel?"

"Can't quite grasp it yet give me a little time! Well, here's to the best chum an expectant father ever had in the bush."

Letter after letter from my brave sweetheart I tore open, only to find out that they all were written before the arrival of our daughter, yes, daughter. I tell myself this again and again! These must have been the letters which accumulated in Fort McMurray and Yellowknife over break-up. They were wonderful. There was no complaint in any of them. Just cheery reports how the days were spent, how wonderful Mrs. B. and Frances and Dr. Terwilleger were taking care of my beloved one....and my daughter Marcia.

Even the best news seemed to take a little time to sink in, otherwise I couldn't have felt a little disappointed not to have received any more details about our new role of Mommy and Daddy yet. I had to keep telling myself, "It's a girl. All is well!" and finally happiness, immense relief, but deep yearning, too, fills my soul. I feel I have to do something, "How about our supper, Art?"

"Supper? Never mind! Didn't want to disturb you. Look at what mother sent us!"

On his bunk Art had spread out the contents of several parcels, cake, cookies, and other delicacies. What a treat! "There's one more here. Has the old man's handwriting on it. Bet you I know what it is. Listen, it gurgles! What a lovely sound! Good old Pop, he never lets me down!"

Out comes the bottle, "Good stuff too! I always says, 'the best is not any too d… good for the likes of us.' Have one!"

"After you, chum!"

"OK. Well, here is to the new baby, the new Mummy and the new Daddy."

"Thank you, Art, in the name of our new little family. And here is to your Mum and Dad. They surely were good to us tonight!"

Then Art said, "Before we do any more toasting, I'd better read my letters. Here is a thick one that is from mother. Listen, Daddy, 'Yesterday I saw Hermann's wife and her lovely baby at the Royal Alec. Both are just fine.'"

"Well, that calls for another drink!" And thus began a memorable celebration. Just the two of us, deep in the northern bush beneath a star-studded sky.

The sun had wandered far and high by the time we rose from our bunks. The light was somewhat hard on our eyes, and the old grey matter seemed to float in its housing.

We both groaned, "Boy, oh, boy," and then as one, "if we had been smart, we'd have saved one for this morning."

"That was some party we had all by ourselves, Art."

"Sure was, and hell, man, what's a party with no hangover!"

Coffee, strong coffee, and lots of it, helped to bring things back to normal, and then, we both started saying it together, "It's about time we find us a mine, baby needs shoes!"

CHAPTER XXXVI
BABY NEEDS SHOES

EASTWARD WE TREKKED, following the low ground. The overburden was heavy, only little rock shows here and there, but we didn't overlook any outcrops. Skirting patches of muskeg and little stagnant potholes, we kept heading east. By and by we got onto higher ground, and into thicker bush. The heat grew oppressive, but the mosquitos didn't seem to mind. They were out by the millions; their buzzing filled the air. Clouds of them were travelling with us. Off and on we stopped to smear some more dope onto our faces and necks, but it's a question whether it's worthwhile. When we stopped plowing through the bush, we gave them a chance to alight, and the dope rather than deter, seemed to attract them. We slapped and swore, but on we went.

The country looked interesting, and we wanted to find that mine. We had travelled a few hours when we came to the brink of a wide trench running north and south. The bottom of the trench was swampy, but the walls of altered and badly crumpled rock look to us like a shear zone, which called for a closer inspection. For a short while we travelled along the foot of the walls, breaking rock, to see if there is any mineral around, but soon we had to give up. The mosquitos, thicker than ever, launched an all-out attack, we could hardly keep our eyes open, the lids were so swollen, our ears were on fire, and we just itched all over.

"Let's get the hell out of here!" we both said.

We scrambled up the far slope. A few hundred yards and another trench was before us. We slid down into it, but this time there was no stopping us. There was no doubt, we have stumbled into the main breeding ground of the stinging pests, clouds of them arose with every step as we practically ran across the bottom, and frantically struggled up the far wall. Our chests heaved, we had never been that hot and bothered before. Sweat soaked and salt encrusted pants and shirts clung to our bodies, rubbing and chafing the burning skin. The longer we stopped, the worse it got.

We decided to cut north onto the higher ground, then swung back to camp along the lake shore. Driven half crazy, more running than walking, we finally reached the higher level. If we had hoped to find relief there, maybe a little cooling breeze, that proved an illusion. A brassy sun was beating down out of a greyish blue sky, there wasn't enough wind to bend a blade of dry grass. Shimmering heat waves were rising above the low ground we had left behind. No use to stop long. As soon as we had regained our wind, we started pushing through the bush again, almost on the double.

At last, we had a glimpse of our lake, and then the shining white spot of our tent. The closer we got, the faster we ran, and when we finally reached the shore near our tent, where the water was shallow, we didn't stop. We just threw off our pack boards and jumped right into the drink. That cooled us off in a hurry. We tore off our clothes then, had a good soap with Lifebuoy, and soon we felt pretty good. A substantial meal and the inevitable pot of coffee did the rest.

"We'll never forget this one, I bet you. That was a hell of a day in the bush!" I said to my chum.

"Sure was. What puzzles me though, what did those bastard mosquitos feed on before we got the crazy idea to give them a free meal?" Art came back. That indeed was quite a question.

One of those short, but violent northern thunderstorms brought some relief from the sweltering heat in the evening. After a good night's sleep, we were ready for some more prospecting. We left the east alone though, crossed the lake, and headed westward on high rocky ground with little overburden and just a few trees here and there.

The formation lay before us like an open book. Old sediments, pushed around and altered by intrusive granites, numerous quartz veins, big and small, and cut through all this, basic dikes showed rusty red. Busily we swung our prospectors' picks, expecting to see mineral, but only occasionally we spotted a few pyrites.

After lunch we travelled south for a short distance and then east again, paralleling our forenoon's journey. We didn't find a mine, but at least it was a day we enjoyed, and after all there was always tomorrow.

Paddling back to camp we saw a big fish jumping near our canoe, and that reminded that we hadn't yet tried our luck at fishing. Visions of a succulent feed of fresh fish arose, and we were not slow digging out our line, spoon, and hooks. The preparations took longer than the catching. About five minutes after we threw the lure overboard and played out some line, I felt a terrific pull and the line slid through my fingers. The big fellow we pulled in could have put up a good fight, but he didn't. So, this wasn't real fisherman's stuff. Just a way of getting supper.

Art skinned and filleted it, and boy, was it ever good, after all the bully-beef and canned sausages! The lake must have been teeming with fish, because from then on, we caught us one for breakfast, lunch, and supper, and it never took more than a few minutes trolling.

"Hello, this camp!" came a shout one morning early, when we had just crawled out of the bedrolls. Two husky, bearded figures alit from a canoe and approached the tent.

"My name is Lundmark," said one.

"Thomson," said the other. We introduced ourselves. The formalities over, we asked them to join us for breakfast. The two didn't hesitate, they made themselves right at home, and that's how we liked it.

"You fellows sure know how to be comfortable," they declared when entering our home.

A good square meal was soon prepared: the inevitable mush, flapjacks, and bacon, and then our coffee pot started rattling the lid. So, we sat and ate and talked. Where are you going? Where have you been? Seen anything interesting?

Our new friends are heading north through Desperation Lake. We told them to help themselves if they ran short of anything, from our cache at the old campsite. The morning wasn't young anymore when we finally parted. We liked our visitors and regretted to see them go. However, every prospector has to follow his own rainbow, and while at it, is better off alone.

Our own search led us westward for several days, we crossed the lake and travelled towards the Beaulieu River. We didn't make a strike, but we had at least a more memorable experience.

Every day we started on our roundtrips somewhat farther north. One day we pulled our canoe ashore at a spot where a ravine cut through the rocky shore. Near the water it was only like a narrow pathway, but farther inland it widened and here we suddenly found ourselves in a forest of giant trees with slim, perfectly straight trunks pointing skyward for a hundred feet or more.

Many had fallen in years gone by, some of them were still quite solid, while others were just a shell crumpling under our feet, but those which still stood, were a rare sight in a country where so far, the biggest trees we had seen were scraggly looking jackpines. Only in that deep sheltered ravine could these proud specimens have survived the northern storms. I was wondering what kind of trees they might be, they looked somewhat like poplars, but the deeply corrugated bark? Art knew the answer. "Here's where I shine. Them's "Ball-McGillia" poplars. That's one thing I learned in school."

Reluctantly we left the cool shadow of the mysterious ravine to travel over the hot rocks of the country beyond, fairly barren of vegetation, and as we found out, barren of mineral too. In all our further travels we never saw such trees again, and it still puzzles me why they should have survived and be confined to that one particular spot. The ground to the west had now been prospected by us, as far as we could wander in a day, and we thought of trying the east again. The next day was a scorcher, memories of the last trip were still too vivid. So, we went to the south where the rushing waters of the Beaulieu River poured into the lake, over a series of rapids and narrow gorges.

Of the prospecting camps we had been told about, we saw none. Partly this, and partly the looks of the country farther south prompted us to plan on a move in that direction. With that in mind, we headed for Desperation Lake the following morning to get the supplies left in the cache. What we found there almost made the trip unnecessary. Near the shore tacked onto a tree, we found a message, "Bears got into your cache. Straightened things up as much as we could but fear they will return. Thompson, Lundmark."

It was only too true, the bears had returned, and a sorry sight greeted our eyes. Hardly anything was left on the cache. Flour and sugar bags had been ripped open; the contents scattered all over the place. Packing cases of canned sausages, bacon and bully beef were smashed, the cans were either bitten or clawed open, and in spite of the jagged edges of the torn metal, most of the cans had been licked as clean as a whistle on the inside. Even the coffee and tobacco tins were badly mauled, but a few were intact. Of the sugar, honey and ham, nothing was left.

All over the place there were unmistakable signs that the unusual diet must have had an accelerating influence on the digestive processes of the robbers. A heavy odor of bear and putrefaction. We didn't stay long. Strangely enough, the tarpaulin which had covered the cache showed only one long rip, and so was worth saving, if little else was.

We were pretty mad at the unwelcome visitors and swore bloody revenge. However, since the planes were flying again, we didn't have to worry about getting more supplies. A plane was due soon, and just in case we weren't in camp when it arrived, we left a list of things needed, the out-going mail and a message for the pilot in a sample bag hanging from the tent flap when we started on another attempt to find out what the mosquitos were defending.

Apparently, they had called on the black flies for reinforcements, because it was another hellish trip with the little black devils crawling into our eyes, ears, and noses, and even under our shirts. Forced onto the higher ground where a light breeze made it more bearable, we never found the answer to our quest, and ever since we remained curious.

Somewhat earlier than usual, we returned to camp at the opportune moment when a Junkers came in for a landing. Ronny and Woods were aboard, looking quite changed from the well shaven appearance of a few months ago, when we had last seen them. Both were sporting beards; Ronny's was a beaut. A reddish blond Van Dyke, which made him look like forty.

"Cripes," exclaimed Art. "For a moment I thought it was Sir Hubert coming to call on us."

"What brings you fellows here?" I asked them. Out of earshot of the pilot, Harry Winnie it was this time, they told us their great news. They had found several showings of free gold, possibly belonging to one lead. Could we give them a hand staking claims? Another party was prospecting not far from them, and they were in a hurry to secure the ground before the other fellows got wind of their find. Sounded logical enough to us, and as we wanted to move anyway, we agreed to help them stake their claims before going to some place of our own choosing again.

CHAPTER XXXVII
ON THE MOVE AGAIN

QUICKLY WE BROKE camp, loaded our paraphernalia into the big Junkers, tied our canoe onto the under-carriage, and off we went. On the way over to Pensive Lake we had the set-up explained to us and made plans as to how to stake the claims in the shortest time.

Pensive Lake was roughly u-shaped, the open end pointing south. The find was on the peninsula thus formed, at the base of which was Ronny's and Woods' camp. Our own temporary camp we established on the northeast side of the lake, and without much delay we began blazing lines and putting in claim-posts. In two days of hard work our posts were in place, and nothing to worry about anymore, we could put in the cross lines in a more leisurely manner, doing some prospecting at the same time. Before we did that, however, we figured it might be a good idea to look beyond the now secure group of claims for possible extensions of the mineralized lead.

To get a line-up Ronny and Woods took us to where they had found the free gold in three or four different places alongside a pronounced shear zone, running roughly south-south-west to north-north-east. This pointed almost exactly to the spot where Art and I had made our temporary camp. The formation there differed from the old, altered sediments in which the gold had been found, and, adding to this my impression gained by looking at the shear zone, I was inclined to think that an extension of the mineralized lead might be found rather to the west than to the east, the north end of the lake possibly being a fault in the formation.

From the higher points of the peninsula, it definitely looked that way to me. Woods, who often uttered some very positive opinions without firm foundations, did a lot of talking against my theory, which I was very anxious to put to the test. Ronny didn't say much, as was his usual manner, but it almost seemed as if he were quieter than ever. Maybe he had had to listen to Woods too much during the last two months, and now even Art's teasing, which used to get him going a little in the past, failed to get a rise out of him.

A few days of Woods grousing and kicking had been enough for us. We were tired by now of his constant squawking about Pat's not showing up, which was the recurring theme of all his talk, and which at that time was rather pointless. We had plenty of work, enough grub, and if we wanted to move, we just had to make arrangements with the next commercial plane that showed up. So, instead of fruitless arguments with a guy whose head had grown too big, we just said, "OK. For one more day we go with you fellows, wherever you think you might pick up your lead. After that, we go on our own. You got your claims staked, that's all we came here for. Now we'll find us a mine of our own."

That day was spent in fruitless search. There was no doubt we were in the granite instead of the old sediments, and my conviction grew firmer that it would be more promising to go west. Looking in that direction when on our way back to camp that evening, I noticed about a mile away from where we stood near the rapids connecting Lower and Upper Pensive Lake a bright reddish-brown band running up a hillside at an angle, its peculiar color enhanced by the setting sun.

"See that over there?" I pointed it out to Art. "That's where we are going tomorrow. That looks interesting to me. From now on, we'll follow our own hunches again."

"OK by me," said Art. "As long as we find us a mine. To hear that other guy talk, you'd think he's the only one who can do that, and I'll bet you it was Ronny who found the first gold. Must ask him some time."

"By the way, Art, who found the gold at Gordon Lake? I meant to ask this for a long time."

"Take it from me, brother, I broke the first rock on that quartz hump, I made the strike. Why do you ask?"

"It was really no question to me, but I've never yet heard anybody giving you credit for that find."

CHAPTER XXXVIII
FREE GOLD

IN HIGH, GOOD spirits, the two of us set out early the following morning for the golden west. Quite a breeze was blowing, and Pensive being a very shallow lake at the north end, there were even some whitecaps to contend with. We snuck around the leeside of a few small islands and safely reached a reed-grown bay which appeared nearest to the red-brown band. From there we made a beeline through a draw with heavy overburden, which led us into burnt-over country, where second growth had made a timid start. Gaunt, burnt trunks of jackpines were still standing up, and the ground was littered with others.

On the ridges, hardly any soil was left, making prospecting an easy matter. The red-brown streak turned out to be a basic dike, cutting the formation at an angle, with very straight, clear cut walls. Under the reddish coat of the weathered surface, we found a dense, granular blackish-grey rock with an occasional speck of pyrite.

Our curiosity soon satisfied, we kept going parallel to the dike, Art a hundred feet to my right. Heading for the crest of a second low ridge, cracking rock while we were going, I gave a wallop to what seemed just a band of discolored rock. Something glinted bright and yellow in the sun.

I bent down to have a better look, then I knew: Only gold looks like that. There it was, shining brightly, a flat nugget about half the size of my small fingernail. I hollered wildly, "Hey, Art! Come over here!"

My chum must have sensed my excitement, he came on the double. "Look at this, Art!" I pointed at the nugget.

Art went down on his knees, had one good look, jumped up, threw his cap in the air, and cried, "You got it, you got it! Free gold!"

There ensued some frantic hitting and picking around our nugget. It turned out to be a badly weathered quartz vein we had found. The rock broke easily, and in no time at all we had quite a few fine nuggets. There was no doubt it was a rich pocket.

"If we only had some dynamite! Boy, oh, boy! This looks rich! Look at this one! Hey, look at this one!" We must have done some excited gabbing. When we cooled off a little, we decided to look how far the vein went. That's where the puzzle started. We ran into a maze of quartz veins, some looked quite normal with well-defined walls, but suddenly they would pinch out or split into numerous small stringers. Others would veer off, almost turn back on themselves, forming drag folds and then continue in the original direction.

It was a most confusing picture, and if someone had told us we were seeing things, we might have believed him. The best description of what the formation looked like was born, "What do you think of her?" Art asked me, after we had run in circles trying to follow the quartz veins.

It popped into my head, and out it came, "Reminds me of Mother making a marble cake!"

"What you say? Marble cake! Boy, could I go for a hunk of that! Frosting an inch thick!"

"If this pans out you will surely have it! Maybe we'll hear the diamond drills soon again."

We went back to our little high-grade pit. We couldn't resist its attraction, and while we eagerly picked away, not without success, we made our plans.

The first thing to do would be to stake some claims, before somebody else got wind of our find. After that, we would move our camp, get some dynamite, and see what's down below in the solid rock. Time was flying while we planned, how fast we didn't even realize, until the shadows grew long, and a very empty feeling in our stomachs told us that it was way past our usual supper time.

We camouflaged our strike and hurried to our canoe. The lake was rougher than ever, but we thought we could make it, going with the wind. Heading for one of the islands that had sheltered us somewhat in the morning, we shipped quite a bit of water. We battled around to the leeside, figuring on going ashore to dump out the water when a voice, somewhat familiar, hailed us from nearby.

"Damnation! It's someone from Oro Plata again!" hissed Art.

"Holy smokes! We don't want to meet anybody now," was my reaction, but there was no getting out of it. The lanky Britisher, the same one we had met before, came straight for us, "I say, there chaps, where are you going? Rather rough the water, y'know."

"Don't say anything," we reminded each other, but that proved entirely superfluous. The man from Oro Plata wanted to do the talking, "I gather you haven't found anything yet. Rather futile, y'know, for a chap to run around in this blasted country, nothing in it, y'know. As the Canadians say, 'barren as a bull's ass.' I say, rather apt description that. Where do you see a greenstone dike, I ask. Nothing but barren granite. Old sediments! Fiddlesticks! Ha, ha! Can't imagine what made them send me here, y'know."

So it went, on and on, the old grouch probably had missed an audience for a long while. That was OK with us. We just sat, and listened, thinking all the while, "we ain't talking, brother," and hoping the wind would calm down somewhat, so we could go home for supper. We were hungry and weary, and ready to take a chance. Maybe it was mental telepathy, maybe our stomachs really growled in an audible manner, in any case, our talkative host suddenly interrupted his praise of the prospecting possibilities in Northern Ontario!

"I say, chaps, a spot of tea might be indicated, you must be famished. Like to offer you a little of our rice pudding. Had our dinner, you know. But always prepare ample. Rather nourishing stuff that, rice pudding. Fancy I'm rather good at preparing it." He surely was an enduring talker. His tea was good, the rice pudding, well, perhaps it made us brave the whitecaps. As soon as we decently could, we said our thanks and farewell, loaded a few rocks into the center of the canoe, and set out on the last, longest leg of our campward journey.

Once out of the shelter of the island the water was rougher than ever. Running before the wind was all right, but when we tried to round the tip of the peninsula to swing towards camp, we were almost swamped. There was nothing left but to swing the canoe at a right angle to the combers again. Fortunately, in that direction the bow pointed straight at a small sandy beach facing the widest open stretch of Pensive Lake.

Swept forward by wind and waves, faster than we had ever travelled in a canoe, the beach came rapidly nearer. Behind us was a lake in a turmoil, in front of us a crashing surf. All we could do was to keep our creaking boat in the right direction. A series of smaller waves were followed by a real big one, and then close to shore the grandaddy of them all boiled up behind us, took us onto its crest, and smacked us down high up on the sand. Hurriedly we jumped out, dragged the canoe higher, before the next big one reached us, and hiked toward camp.

It had been another busy day. From now on there would be lots of them. After a substantial feed, we felt good enough to lay out the claims we were going to stake the following day. That done, we were only too glad to stretch out in our eiderdowns, the howling wind singing us to sleep.

Long before the sun was up, we were crossing the now calm lake, once more keeping as far away from Oro Plata's island as possible. We were in a hurry and needed no witnesses.

By evening we had staked most of the claims, as planned. That had landed us near the channel between Upper and Lower Pensive. Voices drifting up to us from the water interrupted our whittling of the last claim-posts for the day. Casually we strolled over to the rapids, leaving our axes behind. Our friend from the night before and his silent partner were portaging into Lower Pensive. He did the talking while his partner lugged canoe and kicker.

"Gone north today?" we asked very innocently. "Find anything?"

"Find anything, you ask? My dear man, there isn't anything in this country! I have said it before, it's utter futility and sheer nonsense to spend one's time in this field. I shall pull out tomorrow morning."

That was quite all right with us, we didn't try to hold him.

At breakfast, the morning after this encounter, we heard a kicker. Looking out we saw our friend heading southward to try his luck somewhere else, no doubt. Our job for the day was the cutting of the east and west lines of our claims. Halfway through the forenoon Art and I met up at a spot on our westernmost line. Our work was almost finished, one more short line to cut towards Upper Pensive. After that, we could go back to the high-grade pit. While we were having us a smoke, we heard sounds as if somebody were chopping down a tree, not very far to the west.

"If that's somebody staking over there, we'd better put some more 1 and 2 posts on this line," I suggested to Art.

"Yea, we'd better. They don't stake without a reason!" Art replied.

A few minutes later we ran into an elderly fellow who was soon joined by a youngish-one, both carrying axes. He turned out to be a well-known prospector. A strange conversation followed, both parties trying to find out what the other one was up to, without telling too much. It grew more confusing every minute, but I think we got the better of the bargain, by convincing the other party that we had Nos. 1 and 2 posts on the line. In that way, we had extended our ground one more claim length to the west. On that basis, the at times quite heated arguments were settled.

We finished our line to the lake, went west for about 1500 feet, and then south again, without much stopping. We had never staked claims faster. In all likelihood, our competitors didn't lose any time either when gold was at stake. Our southernmost claims enclosed a low, sandy point, a rather desolate-looking spot. It definitely lacked beauty, but other advantages made it an ideal campsite. A long, sandy beach promised good bathing and the exposure to all the winds that might blow would keep mosquitos and flies away. Many charred tree trunks littered the ground, plenty to keep our pots boiling.

The base of the peninsula was quite narrow, which would make it easy to clear out a fireguard. It was a bad summer for fires all over the north the pilots had informed us. Quite frequently the atmosphere was smoky, indicating that some of the bush fires were raging not very far from us. Planning on moving our camp to the selected site, we were now hoping that a plane might drop in soon, and we were happy to see our hopes fulfilled when Harry Winnie brought his ship down near our temporary campsite, the same evening. He brought us long-awaited mail too, lovely letters and even the first pictures of our baby daughter, Marcia. My heart was filled with a strange mixture of joy and sadness.

Harry came back the next day, and in short order we took our outfit across the lake.

Our find and the staking had kept us so busy that we hadn't spent much time on wondering how our other two prospectors were making out or whether Pat would finally show up. It was towards end of June by then. Two discoveries had been made, a great success, but we shouldn't rest on our laurels. The more ground we could cover during the short summer, the greater the chance of finding something really worthwhile, and for that purpose our outfit's airplane was meant to keep us supplied, to pick out likely-looking areas from the air, and to drop us into the handiest lake for a closer inspection.

For these reasons we were looking forward to having Pat and the plane permanently with us. Pat had said something about getting a new plane, a Waco instead of our old trustworthy De Havilland, maybe there had been some delay. When we heard a plane, making an unfamiliar sound, landing at the south end of the lake near Ronny and Woods' camp, we canoed down to have a look. With us we took a Milk of Magnesia bottle, large size filled to the stopper with small specimen of rock, all showing free gold from our high-grade pit.

It was Pat all right, there was a black and orange Waco tied to the shore. But if the plane was different, so was old Pat. Grumpy in great contrast to the high-spirited fellow we used to know. After all, the other fellows must have told them their good news, and here he was kicking about his new plane, the long delay in getting it, and now the damned thing wouldn't get off the water, although a new propeller had been installed to improve the poor performance, and there were other troubles, and he had to get back to Edmonton in a hurry.

When it came to this, I thought it was time to get a few things straightened out, "Why in hell do you want to get back to your troubles? There are no troubles here. You'd better stay with us. We need the plane, there is a hell of a lot of ground to cover yet. We have waited a long time for you."

"There are troubles I can't tell you about, besides the fact that I had to fight that damned crate all the way down from Edmonton. I'll show you tomorrow morning, it just won't get off the water with any load on. This is a fair-sized lake; I couldn't dare drop into a smaller lake. Holy smokes, I don't like it any better than you do."

"OK. So, you get the plane fixed up, and then you come back to stay. Is that it?"

"There are other things!"

"There was a time when nothing would daunt you. You'd better try to regain that old spirit. This is just the right climate for it."

To clinch my argument, I finally brought out our blue bottle, "Have a look! Maybe this will cheer you up!"

"Holy Moses! Where does this come from?"

"From our find! There's lots of it!" chuckled Art.

"Your find? What do you mean? I thought only Ronny and Woods made a strike!"

"This is the first time they hear about our strike," I explained.

"That's what I call spectacular free gold! Where is the vein? How long is the lead, how wide? Did you stake it?" Pat had so many questions to ask all at once.

"It's staked all right. How long and how wide the show is we can't tell you that yet. You come on over tomorrow and we'll show you. It's inland from the northwest corner of the lake. Feel better now?"

Darkness had fallen by then. If we didn't want to sleep on the bare rock, it was time for Art and me to return to our camp, after extracting Pat's promise to come over first thing in the morning.

In the morning the five of us trudged up the well-worn trail to our showing as the rising sun painted all the western hills golden. Our glory hole made the proper impression on our visitors who had no difficulties picking up a few choice specimen of free gold for themselves, but when it came to determining the line of strike, width, and length of the different veins, they were just as much at a loss as we had been.

"Craziest thing I've ever seen!" was Pat's verdict. "You did the prospecting," he turned to Art and me, "what is your conclusion?"

"Only the diamond drill can determine the extent of this thing," I ventured to say. "It doesn't seem likely that the gold appears only in one spot, although we haven't found it anywhere else yet. I'd like to think that this showing widens out the deeper we go. We may be on top of a big body of quartz, and the country rock is, so to say, floating on top of it."

"Seems a novel idea. How are you going to prove it?"

"If you are still determined to go south again, make arrangements in Yellowknife to have some dynamite sent out. For the time being, it might answer a few questions," I had the plan already.

"OK. I'll do that. Need anything else?"

"Lots of things. We got a list in camp, that is for grub. With the powder we need fuse, caps, some steel, a blacksmith's hammer, sledges, a forge. If you'll get those things, we can pick up the grub when we go to Yellowknife for recording."

"Well, I'd better be on my reluctant horse again. I hope to be back again soon. In the meantime, I'll ask Canadian Airways to look after you guys. And keep your noses dry! How about letting me have your samples, Hermann? I want them for the boss, they will do more than all the talking to bring the drills in here."

"If that is so, I'll part with them. Art and I were going to have us a ring made from our own gold, but there's lots more where this came from. Hate to see you go, Pat! Remember, there's no trouble here. Better come back soon!"

No trouble? Hardly had Pat taken to the air again, when Woods started to shoot off his mouth about him; he was just hanging around the beer-parlors in Edmonton, the plane's failure was just an excuse, etc., etc.

We had listened to a lot of that stuff before and grown pretty tired of it. Naturally, we were disappointed too, but we felt that there must be compelling reasons for Pat's behavior but why a fellow, whom Pat had given a big chance should feel he had the right to talk wildly like that about a fellow we held in high esteem we just couldn't see. Furthermore, Woods had only joined our outfit after we had made our first strike, which had assured the continuation of our prospecting activities. Woods had always had that first strike, of which he had played no part, to fall back on. Those were my thoughts, and when the wild talk wouldn't stop, I grew pretty hot under the collar.

"Why in hell don't you quit? Why didn't you open your big mouth when Pat was here? You sure acted as if you were his greatest friend. You better shut up while I am around!" With that I finally put an end to the squawking, but ever afterwards a certain mistrust lingered in my mind. How loyal will a fellow be to his co-workers or to his outfit, if he can't be loyal to someone who was a friend to him?

In the game we were engaged in, you had to expect that the fellows you worked with keep their bargains, or the fruit of your labors would constantly be jeopardized.

Many a mine owner had lost his mine because of careless talk, boasting and plain greed. Were we going to be victims too?

July had come. Shimmering heat was lying over the country. The bush was tinder dry, the moss crumbled into dust under our feet. Lazy breezes from the west carried dense clouds of smoke with them. Somewhere fierce fires were devouring the bush. We got busy and cut and cleared a wide fireguard across the base of our Peninsula. Heat, smoke, and the roar of airplanes from early morning till late at night, that is what I remember of those dogdays at Pensive as Art and I went about prospecting our claims closely, and with hand steel and dynamite tried to get an answer to the big question, how big is our show?

Not unexpectedly the sound of our blasting soon brought visitors to our camp. Strangers and fellows we had met before dropped in, and to all was extended northern hospitality. This made our supplies dwindle faster than ever, and as we had to record our claims anyway, we went to see Ronny and Woods, figuring that we all could go at the same time, and tell the next pilot to have our bunch picked up.

That was quite all right with Ronny, but for somewhat mysterious reasons, Woods wouldn't agree. When, instead of talking about the business on hand he started one of his tirades against Pat again, it didn't take us long to make up our minds to go back to our own peaceful camp. Just then a plane arrived, piloted by Con Farrell, who never let you guess about his feelings. Something had gone wrong somewhere. Steam had to be blown off, and it was done in a familiar forceful manner.

We had all had enough, and an argument broke out amongst us. Should we all go to Yellowknife together and when? It was too hot for civilized conversation and it was instead an argument.

After that, Art and I canoed home. Two days later a plane came to whisk us to Yellowknife. The other two fellows had gone the day before.

CHAPTER XXXIX

YELLOWKNIFE

IT WAS MY first visit to the new boomtown. Art was my guide, that is, in the beginning, while his resolution to stay sober was still strong.

"Stick with me, Hermann, or I might go on a toot, and I've no intention to do that. But one never can tell, can one?"

"OK Art. We'll have us a drink when our business is finished. Where do we go first?"

"Let's go up to Vic's to get us a room."

Vic Ingraham's Yellowknife Hotel was quite an imposing structure, measured on the shacks and tents that made up most of the settlement. Having been an active prospector himself, before a narrow escape at Hottah Lake injured him, Vic heartily welcomed us.

"Sure, I will give you a room. Any guy who comes here direct from the bush gets a room, even if I have to throw out a few travelling salesmen. Just dump your packs. You are welcome."

"Quite a guy, that Vic!" said Art, "Ever hear his story?"

"Jack Lorne told me about it. He was one of the party looking for radium at Hottah Lake. Must have been grim."

"What's next on the program?"

"Let's get some claim forms from the Mounties. After that, we go to Mike's to order our supplies."

The RCMP Post was housed in a neat, solid log cabin. Inside, the combined living quarters and offices were cramped, but about it all there was an air of efficiency and orderliness, even to the jail, a stout iron cage in one corner. Through the bars though, a strange sight met our eyes, a few cases of the best: Haig & Haig, White Horse, Buchanan, etc. A sight to make the strongest waver.

After we left, Art told me about the system of liquor control in Yellowknife: only the white men can have it shipped into the Northwest Territories on a permit. Vic was sort of "official bootlegger". He collected the permits and ordered the stuff. The shipment would go to the RCMP who doled it out a case at a time to Vic, who in turn supplied his customers. Periodically a Mountie would make an inspection tour. If no drunks were lying around, no noisy brawls broke the peace of the settlement, Vic could send his factotum with a pack board to get another case. It seemed a very sensible solution to a perplexing situation.

"Mike's" at the other end of the settlement, combined grocery, dry goods, and hardware store, also the Post Office, and was a busy, crowded place that morning. Whenever we had been in need of anything, we had sent a list of things to Mike's Store. The quick service and the quality of the goods had been very satisfactory, but what had pleased us especially, was a bunch of unsold magazines thrown in with the supplies ordered.

On the crowded shelves were many good things we had done without for a long time. We were kids in a toy store. Many a packing case was filled before we were through. On top of the last one nearly went a typewriter. Mike, never missed a chance to make a deal, and almost talked me into buying it. However, I wasn't any too flush, and after all, I had it in my mind to ask one of the pilots going to Edmonton to send some roses to my two sweethearts in Edmonton.

The Wildcat Café was our next destination. It was time to have some lunch. The place was filled, but off to one corner we managed to find seats. It was a good vantage point to study the motley crowd, while we were waiting for the grub, we hoped the queenly waitress who lorded it over this bunch of he-men, would bring. She was in no hurry. A group of pilots were the main attraction as far as she was concerned. Oh well, those guys had glamour, we were just a couple of prospectors.

A few faces were familiar, fellows we had met in the bush, but the majority were strangers, some of them altogether too smooth-looking to be taken for guys who spend most of their time cracking rocks and fighting flies. It was a little reminiscent of those Western pictures, where the well-dressed city slicker takes the trusting cowhands to the cleaners, when we noticed one of these characters, in whispered conversation with a grizzled old bush rat. Maybe the old fellow was just trying to get himself a grub stake with a tall story, of a bonanza to end all bonanzas.

Look! There it came; the old fellow unwrapped something, must be a piece of high-grade! It wouldn't be a mining camp without that.

"What else can you show me in this town?" I asked Art when we left the Wildcat.

"Don't know. The place has changed quite a bit since last year. Bet you anything though, they must have a gambling joint. Oscar Lanny was running it last summer. You know, the same guy who had one in Goldfields."

"Yea. The guy they call a square shooter because he buys them a meal after they have been cleaned out."

It didn't take long to find the joint. On our way we passed a neat looking frame house, freshly painted. "Look at that, Art. They are beginning to put up some nice houses."

"Not bad. If I am not mistaken, that belongs to the best-known whore in town. Sure, that's her, hanging out the laundry. Prosperous looking, ain't it?"

The gambling joint was going full blast. It was Lanny's outfit all right. A short stubby fellow, with quick darting eyes and a great air of good fellowship moved from table to table. Welcome, boys! Care to sit in?" he boomed when he caught sight of us.

"Not today, thanks, we are out of funds. How are things?"

"Quiet. Just before payday. Tomorrow we'll have some good games."

One of his henchmen came up. He needed some change. A cattle-buyer's roll was nothing compared to the thing Oscar nonchalantly dragged from his pocket. He sure had struck pay dirt! Our sightseeing over, we took a short cut to the hotel where I was going to fix up our claim-papers. The short cut turned out to be the long way home. From a tent a familiar voice hailed us, and thus we got off the crooked and narrow path. Our old friend Punk had seen us passing by. Ronny, Woods, and another fellow who was well-advanced in the process of getting oiled were also on hand. Punk introduced a crock, "Come on, have one! For old times' sake!"

Art and I had the best intentions not to touch the stuff before all our business was complete, but meeting Punk was of course a special occasion, so, "OK. Just one. Here's to the good old days!" we said, tipped her up and wanted to be on our way.

"Ah, come on! Have another one," Woods now took over. It didn't take much persuading to see little Art hoisting the crock again, and that was too bad because after the third one, it was of no use any more to remind my chum of our agreement to stay sober. I wandered on alone.

"Where's your sidekick?" asked Vic when I got to the hotel. "Pretty far gone by now, I'm afraid. Somebody put up a stronger argument than I."

After I finished my homework, supper at the Wildcat Café seemed a good idea. I was somewhat late, and my hope of meeting up with some friends to spend the evening with was in vain. A little lonely and at loose ends, I made another round of the town.

Where the boardwalk skirts a rocky promontory, something sailed over my head and plunked into the water. Somebody hollered from high above, "Watch out below!"

Again, a missile came sailing down, this time hitting a big rock, and breaking into a thousand pieces. A beer bottle. Must be a party up there. On top of the hill, near the radio shack I found quite a gang of pilots enjoying some cool beer and the cool evening breeze. But the welcome was warm. To my great joy friend Wop was there too. He had just come from the south only a few days ago, he had seen Lotti and the baby! Here was the one who would get the roses for my beloved ones. That off my chest, I asked endless questions about the girls. Wop understood.

Some more empties had accumulated, and the contest got under way again. Whoever hit the big rock way down below in the water was the winner and earned another beer. Soon the supply was exhausted, and then we just sat and talked and looked out over the brilliant expanse of Great Slave Lake mirroring the setting sun.

Vic was still at his post in the small lobby of his hotel when I returned. That was the night when I first heard the full story of that hair-raising trip across Hottah Lake. The trip which left such deep scars, when everything had gone wrong, but in the end, human endurance, courage, and comradeship had won out after all. And here was Vic, undismayed, a successful man, a man who believed in the North. Respected by even the toughest hard rock miner, he kept order and peace in his hotel, and should somebody in his cups dare to break the rules of the house, it would be just too bad. There's a handy sawed-off billiard cue under Vic's counter.

Suffering from a severe headache, somewhat crestfallen and shaky on his underpinnings, Art showed up next morning.

"Oh, boy, oh, boy! What a night! What a headache, little Art sure got his load this time! Why did you leave me with those guys?"

"That's what you say now. Last night you thought me a pretty poor sport!"

"OK! OK! Don't rub it in. My poor head. Must have been lousy moonshine. What's up this morning?"

"Got to see the Mounties. The papers are ready!"

"What a life! Us poor bastards fighting the flies in the bloody bush, finding mines for a bunch of smart guys on Bay Street. Bet you they drink better stuff than what I had last night!"

Our business with the Mounties finished, we two staggered along the boardwalk towards the Wildcat once more. Art still had quite a time keeping on an even keel, but he was beginning to feel better and bubbled over with cynically humorous remarks about the world, the North, himself and so on, and hitting on the truth more often than he perhaps realized.

At Mike's, after lunch, we checked the shipment of stuff that had been readied. Once again, a pile of recent magazines was thrown into the bargain, and then we carted it all to the dock for the return trip to Pensive. We were glad to go.

CHAPTER XL
ITCHY FEET

NOW THAT OUR claims were properly and legally safeguarded, we felt that we should move to another location. It was getting too crowded around Pensive Lake, especially after that trip to Yellowknife. We had a strong suspicion that someone had talked too much, possibly on purpose.

To the south of our claims some friends of Woods had staked; to the west, Lambert Turcot was busy on his claims, and on upper Pensive, Dome Mines were putting up a camp. We were properly bottled up. With steel and dynamite we had gone to work on our high-grade show, even a few shots seemed to show a widening at depth of the quartz lead, but hand steel work was tedious, and slow, a waste of valuable time.

A report of all this had gone out to Pat, telling him we would make arrangements for moving to a new district. The answer came quickly, "Shall be there shortly. Wait." and wait we did, not inactively, but regretting to see the summer days pass. The worst of the fly and mosquito plague was over, gone was the oppressive heat that had bothered us. In short, it was ideal prospecting weather. Some deer flies and bulldogs were still around, but they only pestered us when we took our evening dip. The bulldogs are the big fellows that sort of crash land on you, and as the story goes, "tear a hunk out of your hide and fly to the next tree to chew it up."

As long as we were under water, we were all right, but as soon as we got out, there they were. We must have been a strange sight, jumping and running around in our birthday suits swatting and swearing while struggling into our pants and shirts.

Planes were roaring overhead almost all day, and many a kicker churned the waters of Pensive Lake. There were visitors aplenty. Jim McEvoy and Terry Mahoney opened the procession, coming down from their camp at Upper Pensive. Then a brand-new, shiny plane, bearing the grandiose legend *Territories Exploration* on its fuselage, ground onto the sandy beach in front of our camp. They all wanted to see our show, but in that we weren't any too encouraging. If they made a strike, would they be anxious to show it to us? We doubted it.

When Lambert Turcot showed up one day, we made a deal with him, "We show you our stuff, if you show us yours." When he agreed, we set out for the high-grade, which was on the third ridge going inland from our landing place in the little reed-filled bay. We had almost come to the summit of the first ridge, when a totally unexpected sight made all three of us plunk down on our bellies. In the draw beyond, maybe thirty feet from us, peacefully browsing, stood the most magnificent bull-moose we had ever seen, his sleek coat glistening in the evening sun.

Art whispered, "I'll get the gun," and snuck back towards the canoe. The question was, would he return to take a shot? Or would the moose be gone when he returned?

The mighty animal had no inkling of our presence so far. He kept on feeding, only once in a while lifting that strange massive head of his with the giant antlers, turning it slowly, long nose and big nostrils quivering slightly, sniffing the wind. Nothing suspicious around as far as he could make out. Fascinated by the strange, primeval sight, we watched for about half an hour. The snap of a broken twig, coming from behind us, betrayed Art's return. With a snort, up came the big head, hooves pounding the rock he wheeled around. For few seconds he stood, facing in our direction, every muscle aquiver, wound up for immediate action, then, with his legs gathered closely, he went way down on his haunches, before launching into a terrific leap and quarter-turn away from us. Crashing straight through the bush he went east at an incredible speed. Art came running with his rifle at the ready. He was too late. By the time he caught sight of the big fellow, he was too far for a fateful shot.

We other two weren't sorry, and even Art's disappointment soon evaporated.

"Damn funny, though, just the same. For weeks on end, you pack a rifle. You never see a thing, not even a lousy spruce-hen. First day you leave the damn thing home, you bump into a great, big, fat moose. One just never can tell, or can one?"

Old Lambert liked our showing and invited us urgently to come and see him. He was a square shooter, and both of us enjoyed his company. Very taciturn at first, he soon gave the impression that he felt at home with us. It wasn't long before he told us many a good yarn about his prospecting trips, how he hoped to make a real good strike, maybe the last one was OK, but before the diamond drills got in, you just couldn't be sure. Once he made that big strike, he could live in Montreal. He had a sister there, and two most charming nieces, and about them he grew quite sentimental. His heart was lonely. So, we did a little dreaming together that night of the rich strike, and then, maybe, we could all go to Montreal together, to have a whale of a time.

Hours spent in speculation passed quickly, and it wasn't long after old Lambert pushed off in his canoe that we crept into our bedrolls. An earthquake couldn't have awakened us more rudely than a plane missing the ridge pole of our tent. The thunder of the engine tore us out of our deepest sleep. "That's Pat, trying to blow us out of bed with his slipstream!" was our first thought as we slid into our pants.

Alas, it wasn't Pat, but a Mackenzie plane, that came taxiing towards camp, then got stuck on a sand bar. The pilot was the first to wade ashore. "I'm sorry to do this to you fellows, but I had to get rid of my load. Hope they won't be too much trouble," he said.

"What are you talking about?" we wondered.

"You'll see in a minute," he replied, and then we saw three figures unsteadily emerging from the cabin and stagger towards terra firma. "There you are! Three pie-eyed pioneers. They insisted on being taken out of Yellowknife. Maybe they were right!"

Our inebriated visitors approached, one was very polite and apologetic, the other was rather morose and cranky, and the third was beyond caring. He just wanted to lie down and sleep. Before we bedded the tired trio in our storage tent, we helped to swing the plane around. Off it raced into the silvery path of the moonlight to hurry back to Yellowknife. Our guests were soon snoring it off in unison. Far off a loon was crying, and once more we too surrendered to Morpheus' arms.

Strong, black coffee in giant quantities was the mainstay of our guests' breakfast the following morning. They were pleasant enough guys, who had gotten their directions somewhat mixed up. The Dome Mines' camp, lately come into existence at Upper Pensive had been their destination, and that way they headed after we parted with assurances of mutual esteem.

"We sure had plenty of visitors lately," I remarked to Art, as we were washing every last one of our dishes. "How about us going visiting, before any more arrive?"

"Good idea! I'd like to see the Dome set-up, and we always wanted to have a closer look at that high range to the north."

"Let's go tomorrow then. I'd better make some bannock. All these guests have cleaned us out."

Often, we had looked toward the north, where perhaps six or seven miles from our camp a high chain of hills, running from east to west loomed invitingly. "Someday we'll go there," we had often said wherever we had found ourselves in our northern wanderings, the highest of the farthest mountain lured us to its crest.

We rose with the sun, and before it got hot, we had portaged twice, and reached Upper Pensive. There was no sign of life at Lambert's camp, so we kept going towards the Dome outfit. Almost there, we were hailed from the shore by a familiar voice. Coming close we saw it was Terry Mahoney.

Something must have happened to get him up that early, we figured. True enough, a bear had paid a visit to his camp at sunrise, ripping the tent just about where Terry had his nose sticking out of the sleeping bag.

"Damned unpleasant awakening, I assure you. Thought I had a nightmare at first. The unmitigated gall of that thing, blowing down my neck!" Terry reported. "Where are you fellows going?"

"Thought we would have a look at the Dome works. You have seen the show?"

"We'll go along with you to see how far down they got." Quite a few prospectors had been attracted by the news that Dome had brought in diamond drills. At the camp we met up with our old friend Steve of Moore Lake days. He was currently with an outfit Vic Ingraham and his brother were interested in. There also was Jim Blaiswell, so we could be sure to hear some good tall stories. He hadn't earned his name "the biggest liar in the North country" for nothing.

At the cookhouse entrance a fair-sized black bear was rummaging in the garbage cans. Too engrossed in his own treasure hunt he scarcely paid any attention when the whole bunch of us filed by.

Two small X-Ray drills were churning away in a formation somewhat similar to our own show. They hadn't reached any great depth yet. The going was slow, the runner told us. "Damn rock is harder than the hubs of hell." We had found that out with our hand steel work. In the quartz we hadn't gotten anywhere, but even in the much-altered wall rock the going was tough.

Jim Blaiswell did most of the talking, as was his custom, whenever he had an audience. Somebody had goaded him into recounting one of his hair-raising adventures. He was going strong. Once more he was scrambling up a tree, with a pack of starving wolves snarling at his heels. Could that guy talk! While he was holding forth, he was the only one who didn't notice that the bear from the cookhouse was ambling towards our group. Closer and closer he came, driven no doubt only by curiosity.

Now he was only a few feet away. Jim was just telling us how close the wolves got to him, "I could feel their breath, I tell you."

That's when one fellow drawled, "Just about as close as now, I guess. Jim why don't you look behind you?"

One look, one shriek! The bear bounced off in one direction, the storyteller in the other, while the rest of the gang laughed their heads off. After that, Art and I got under way again, the high hills were still beckoning.

An unusual discovery, massive tangles of raspberries growing in sandy ground along the foot of the granite chain, kept us from climbing up to the crest. The berries were ripe and sweet, good-sized too. Having missed a treat like that for years, we ate our fill with the greatest enjoyment, not caring how the time flew.

"We have to come back soon," we decided when we realized that the sun was sinking low. On our way up we had noticed a small clear stream meandering through the sand flats, stretching along the foot of the high range.

We wondered how far it might be navigable as we turned our canoe into it. It was like entering a different world, so unlike our usual harsh surroundings of naked rock, stunted or half-burnt trees, this friendly, unhurried little stream with its grassy banks, here and there extending into fresh green meadows fringed living shrubs and trees, it was like an oasis of loveliness and peace.

This was the second time we had so unexpectedly come upon a scene of pastoral beauty, and its charm captivated us completely. Quickly we pushed along, drinking in the lush picture of our surroundings, the only sound an occasional thump of our paddles against the gunwale of our canoe.

We could clearly see the sandy bottom, and once in a while a fish would dart away, frightened by our invasion. Slowly we rounded bend after bend, our keel just a few inches above the sand. We weren't the only ones who liked the little stream. Rounding another bend, the last one, we had decided. Art suddenly turned around, put his finger to his lips, after digging in his paddle to stop the canoe. A little way farther upstream a bear mother showed her two cubs how to fish. She was so intent on her demonstration that she hadn't noticed us at all. Again, and again, she made lightning quick scooping motions with her forepaw, and suddenly we saw something silvery flash through the air. The old girl knew her stuff. Fascinated, we watched for a while, then slowly drifted backward, so as not to disturb them. Terry hailed us again as we were paddling past his camp, "Come on, fellows! Have supper with us."

Terry was always good company. We gladly accepted his invitation. He was an interesting character. About the supper we were slightly less optimistic, the camp was definitely a "here today and gone tomorrow" affair, that much we had seen in the morning, and that was probably all it was meant to be. A pot full of Mulligan stew was simmering over an open fire.

"Won't be enough for all of us," said Terry, "Tell you what I'll do. Make us some good rice pudding!"

It turned out to be something special. Into a big pot of boiling water Terry poured freely out of a bag of rice. A few minutes later it began to mushroom over the rim. Down to the lake went the cook, where he scooped off some of the surplus rice, then let some more water run into the pot. Back to the fire it went. Up came the mushroom again. Down to the lake, back to the fire, and again the mushroom. For about half an hour our host was the busiest man in the north country, travelling with his rice pot between fire and water, until finally things were brought into balance, and the raisins could be thrown in.

Full of stew and rice pudding, we then squatted around the dying fire, and pow-wowed, and soon there came in the inevitable question, "What's the matter with Pat? Why isn't he with you fellows?" and to that question we would have liked an answer very much ourselves.

CHAPTER XLI
FALL OF THE YEAR

WE WERE IN the month of August, with the days getting considerably shorter. Another six weeks and the season would draw to a close. Would there be work for the Winter? We were still hoping, although we knew it would be high time to make preparations, and so far, not even the big boss had come to look at our finds.

Once more Art and I decided to make a move on our own hook. As we didn't have the plane to spot likely looking ground from the air, we picked an irregular shaped lake about ten miles, as the crow flies, to the north-west from our current camp. While we studied our maps, we were surprised by another party of visitors, Hank, old man Taylor and two other fellows. Seeing our maps, the first question naturally was, "You fellows going to move?"

"Sure, we'll move! It's getting too damn crowded around here," Art answered for both of us.

"Where are you going?" That's definitely a fairly impertinent question to ask amongst prospectors.

I couldn't resist the temptation to say, "Tell you what. We don't know where we want to go. We'll take a chance. I take this here piece of high-grade and throw it at the map. Wherever it lands we'll go."

It was a good thing they didn't believe me, because the high-grade landed right on the chosen lake. We never got there though. A mail plane came in that same night, bringing news from Pat. Within a week he would definitely arrive, and he knew where we should go next. That didn't improve our humor. How many times had we heard that story, all through the summer, which was almost over.

"To hell with that business. If he isn't here within the next three days, we'll move. In the meantime, we collect a little more high-grade for our rings, and then have a look at Lambert's show."

Around noon the next day, after a profitable spell of cracking rock in the high-grade pit, we trekked over to our neighbor's camp. Unfortunately, he was not alone. Two fellows, more like city slickers than northerners, sporting unsoiled breeks and brand-new high hunting boots, looked definitely out of place at the old sourdough's camp. They talked loud and long, quite convinced of their own unfailing judgment. Our friend didn't say much, but puttered silently around his tin stove, as he prepared the lunch. When the questions of the city fellows, concerning our activities grew somewhat too inquisitive he winked at us slyly, probably recalling our first meeting, when we played hide and seek along the claim lines.

We had come at the wrong time. Figuring that some deal between the smart guys and Lambert was probably in the making, we hit the homeward trail.

Our confidence in Pat's showing up in time wasn't very strong, and determined as we were to move, we went to Ronny's and Woods' camp the following day to inform them of our plans. Again, we were treated to Woods long tirades against Pat, only now they were more violent than ever, "What did the little bastard do in Edmonton?" He, Woods, wasn't going to be fooled any longer. He would look out for himself, and so it went on and on.

"When Pat finally comes, I suppose you are going to tell him, all that," I finally cut his monologue short. "Are you going to tell him too about this business of looking out for yourself? I am under the impression you did that all right." That did the trick. His silence was proof enough that he had made deals of his own, as I had long suspected when it got so crowded.

Ronny didn't say anything. Never a great talker, he was now depressingly silent. The only thing I remember hearing him say was, "If you fellows care to eat, I've got some Mulligan here," and a little later, "Maybe there will be a war. Then our worries will be over, or … maybe they just start. Who knows?"

Yes, who knew? There had been alarming reports in the somewhat aged newspapers, which had reached us via Mike's from Yellowknife. But in the bush, the noise from the outside world seemed far away.

Maybe we had been too preoccupied with our own problems, the small daily ones, and the one big issue, which was uppermost in our minds, endlessly discussed; we had explored scarcely a pinpoint of Canada's Northland. Hundreds of thousands of square miles of favorable ground were still waiting to divulge their hidden treasures. We wanted to follow the call of those vast regions, as yet only seen in part by a few trappers and Indigenous people, and a handful of people who had flown over them in airplanes. To keep going we certainly didn't want war, and so wishful thinking may have had its share in our underestimation of the danger. Our deep urge to keep on exploring explained, too, our disappointment as far as the summer's work was concerned.

True, we had found a promising show of free gold, but we had been hampered in our movements. Some of the far hills had been reached by canoe and long treks overland, but beyond, there were so many others. After our first discovery, everything that followed had been an anticlimax, and now summer drew to an end, the nights were growing chilly.

When Pat actually arrived, just before the three days were over, it couldn't make much difference anymore, and Art and I didn't leave any doubts in his mind as to the way we felt about the whole thing. Even all the disturbing news about the world situation, excitedly told, didn't move us much. "What were you trying to do, save the situation from Edmonton?" we asked him, but the usual Irish reaction didn't explode this time.

Instead, we heard long explanations, which didn't help at all. Something had gone out of an old chum, a good blow up would have been much better, and we all would have known where we stood. When the Irishman stopped fighting, then there must have been something wrong, but whatever it was we didn't find out, and didn't much care anymore either.

As it was, Pat's coming only accentuated the anticlimax, and with very little enthusiasm did we listen to descriptions of those promising looking spots Pat said he had seen from the air. The next two moves turned out to be complete flops. Somewhat south and west from Pensive Lake we camped on the east shore of a nameless lake after our first move.

Going west and north we ran into barren granite wherever we went, and eastward the ground was staked solid. On top of that, we soon found out that we had many close neighbors, among them the Despard Brothers, Tom Payne and his gang, Pete Davidson, Lorne, and even good old Pete Lauder.

That lake was a regular meeting place for prospectors. After we had paid our calls, and had the whole situation explained to us by Tom Payne, we were quite ready to move to the second spot chosen by Pat. As we approached it by air, I couldn't see anything interesting about it, it seemed to be all solid granite below. Burnt over years ago, the reddish hills were bleak and forbidding, and that impression wasn't tempered much by the crystal-clear water of the long lake. The clearness of the water was only proof the lack of topsoil, hardly any vegetation had as yet managed to get a foothold again. There was something grim and hopeless about those naked ranges of hard rock surrounding us, and we were indeed sorry not to have insisted on going to the lake of our own choice.

To add to our misery, Pat one day brought old Reg along, and left him with us. He acted as if he had been insulted and squawked from morning till night. We figured the guy was homesick for his Edmonton flame, and that we could understand, but why should he try to make our lives miserable, unless he needed fellow sufferers? Only once we talked him into going with us on a day's trek through the wilderness. He whined all day, and towards evening he lamented so much about his sore feet that we became worried. What did he expect us to do? Carry him home?

The next day Pat appeared with Maud, which was a surprise. We didn't waste any time, started packing immediately. "What's the matter with you guys? You don't like it here?" Pat started to wonder.

"Look! Brother! We are getting out of here. We are going to a place we pick ourselves and take your goddamned grease monkey with you and keep him. We mean it. The whole season has been fogged up by delays and waiting, for what reasons we don't know. We are pretty sick and tired of it, and one thing we don't want to hear any more about, is the way you fellows suffered in Edmonton, nursing a crate that wouldn't fly or something," my opinion was shared.

"Holy cats! You must be mad! Are you bushed?"

"No, neither the one nor the other, just outspoken. Are you going to take us out of here, or are you going to send a plane so we can move?"

"All right, all right. If it's no good here, you can move, of course. Here I set you down on a contact zone, and you say it's no good. Why?"

"The contact you talk about is about a mile to the east from here, and that ground has been staked solid. By the looks of the posts claim lines, a bunch of Peace River farmers must have been at work there, but it's staked just the same. We are not interested in prospecting on somebody else's ground. There is a whole continent we can explore, without somebody sitting on our tails and vice versa."

"OK. I'll make arrangements for a plane as soon as I get back to Yellowknife. I am going to see the other two fellows first though. I don't expect to hear so many complaints from them."

"I don't either. We heard them all summer."

"What do you mean?" Pat asked stumped.

"Exactly what I said. However, you'll never know."

"Maybe I do. Where are you going from here?"

"We want to get our canoe from Pensive, and we'll pick a fairly large lake from where we can cover lots of ground. There are only a few weeks left before the snow starts flying. We'll let you know where we end up."

"OK. Hope to find you in a better mood when I get back!"

"Stick around. It might improve," I said.

"Sorry. But I have to go! So long and keep your noses dry!"

"No foam around here where we could stick them in. So long! God bless Ireland! She needs it!"

"Cripes!" said Pat and he was gone.

Art and I were alone again, and we liked it, although we did some wondering. What had come over our fighting Irishman? It was the first time he seemed to be on the defensive altogether, and we didn't like it a bit.

Mert Wales came the next morning to move us. Our spirits soared when we saw the Junkers racing towards our five-day camp.

"Where to fellows? Your boss told me to look after you. He left for the south!"

"For crying out loud. Pensive first to get our canoe, then to this lake here!" We pointed it out on our map, but Mert said it was no go. He had tried to get in there before with a party but couldn't make it. Too many reefs. That was that.

"All right. How about the south end of Hearne Lake? Anybody in there now?"

"Not as far as I know. I'll circle a few times to make sure."

"OK. Let her rip."

An hour later the big Hearne Lake was below us. Flying low we surveyed the shoreline. No tents could be seen. So, I signaled to the pilot to take her down.

"Don't forget to take us out before freeze-up," we told Mert after unloading our outfit near the south end of the lake. "It seems we are on our own. By the way, who wanted to be taken into that other lake?"

"Must be friends of yours, they are friends of Woods anyway."

"That's taking a lot for granted. Thanks for the information. That's what I expected to hear."

CHAPTER XLII
LAST CAMP OF THE SEASON

AS SOON AS Mert had left, we went to work putting up our tent, in the proper manner too, log walls and all, as we fully expected to spend the balance of the season in that spot.

By the end of September, winter wouldn't be very far off, and here it was the ninth day of the month. The nights were getting cool, although the days continued to be warm and clear, giving seemingly unlimited vision when we gained the heights on our searching wanderings. Our new camp site, though chosen in a hurry, was a lovely, sheltered spot close to the south end of the lake, somehow very similar to our first camp at Desperation Lake.

Our spirits had risen. Action, and the pleasant aspect our new surroundings proved a tonic, as Art said, "By golly, this looks different from those last Godforsaken stops right in the middle of the bloody granite. Maybe we'll find us another mine yet. Would only prove though how much time we wasted! I feel good. Full of ambition. Maybe I'd better lie down for a while. Next stop Edmonton and the bright lights! Jasper Avenue here we come. I'll sure need all my strength. It will be a long hard winter, struggling from one beer parlor to the other. At least as long as the old mazuma holds out." We were always thinking about whether or not the money would last long enough while prospecting.

So, Art had his dreams, and I had mine. Soon now, after the longest months of my life I'd be with my loved ones. And there would be the wonder of our baby daughter to behold. Pictures had come to me, and letters had told of the growing and unfolding of body and mind of a healthy infant, who was our daughter Marcia! Yet, until I could take the little one in my arms, it all seemed somewhat unreal and far away.

"It will be quite a homecoming for you, Hermann," Art said. "I'll bet that's what you were dreaming about just now! Come on! Let's get going. Baby needs shoes."

"You said it! Here we go once more, looking for the elusive stuff. Cold yellow metal that pays for the warm comforts of life."

The rather low and barren ground stretching inland from the west shore received our first attention, but although the formation, old, altered sediments, was right, even closest investigation showed few signs of mineralization. Day after day we paddled farther northward, to go inland in a westerly direction. In that manner we finally reached the north end of the long lake where newer granites had replaced the old sediments completely. The well-defined contact zone was scrutinized with special care. However, it proved to be barren of any worthwhile mineral occurrences.

"Tomorrow, we'll look at the east shore!" we said, but when morning came, I was awakened by the wild flapping of our tent. A cool stiff breeze was whistling from the north, driving whitecaps crashing against the shore.

Once out of my bedroll, the icy edge of the north wind sent cold shivers down my spine, and I hurried to get the fire going. Good hot coffee would hit the spot. Art stuck his nose out of the sleeping bag when the aroma of java and bacon on the fry began to fill our tent.

"What's up? Visitors?"

"Yea. Visitor from the north. Old man winter's first messenger!"

"Snow?"

"Not yet. Can't be far though by the feel of this breeze."

"Holy mackinaw! What a country! I'd better go back to sleep!" Art said, reluctant to get up.

"Ah come on. It's not a bad day, no flies! We'll make a trip to the south. Can't use the canoe. Lake is too rough. Here, have some coffee. You'll feel better after that."

"Breakfast in bed. I'll be damned!"

"Don't tell that story in Edmonton. They'll think this is a picnic for prospectors."

Well heated from the inside, we battled our way southward through a thick tangle of undergrowth. When we broke into the open, we were facing a big sandy bluff. Fine yellowish sand was slipping away under our feet as we scrambled up to the brink. And there we stood and gazed with unbelieving eyes, like a beautiful mirage the most magnificent park-like scenery lay before us. From the sandy, needle strewn plateau giant spruce trees, straight as a needle, pointed skyward. Some stood alone in solitary splendor, others had younger, eager-looking satellites huddled around their bases. Even the jackpines seemed to grow straighter and bigger than we had ever seen them. The morning sun was painting long bands of shadows across the sand, and it would have been a somber picture hadn't it been for clumps of birches, with leaves turning to gold and their white bark shining.

Slowly and wonderingly, we wandered along, strongly feeling the strange sensation of walking as on a carpet, without having to watch where next to put our feet. Once in a while a chipmunk would scold us, then it scampered away no doubt resenting our intrusion. Those bright little fellows were the only living beings we saw, but countless tracks crisscrossing the sand told us that this was a favorite stamping ground of many animals. There were the big round impressions bears had left behind, the sharp cut spoor of caribou, and the delicate tracings countless birds had drawn in the sand.

Our eyes drank their fill of the enchanted forest as we silently wandered on, unhurried, and the quest for gold forgotten for the while. After we had gone southward a good mile at least, we found that we had traversed our "park", as from then on, we came to call the place. Gently the ground dropped towards a swamp, thickly grown with willows and birches, and beyond we glimpsed the blue waters of the next lake.

"What a place to build a camp," we both said, when we rested for a while on the warm soft sand at the edge of the park, looking dreamily back to where Hearne Lake showed as just a speck of blue, melting into the shining skies where fleecy white clouds were sailing gently. It wouldn't have surprised us much if suddenly a few gnomes, or even a few red coated huntsmen had appeared on the scene which seemed so like a fairy-tale in its loveliness.

Thus, we dreamed and smoked for a while, and wondered how this strange, beautiful place had come into being.

"Let's go and see how big this little bit of paradise really is," I suggested to Art, who, as always, was willing to fall in with my plan, just as I always had an open ear for his ideas, on what to do next. That way it had been with us all the time we were alone together, and to the day of writing, the comradeship of those long months is one of my most cherished memories.

Looking east from where we found ourselves, we saw, not far off, rocky hills marking the boundary of our park, but westward, how far did it stretch? We walked a long way, perhaps two miles, before we got onto wavelike sandy ridges which slowly rose to the foot of the next rocky range, where the bush grew into a dense jungle.

"We'd better look at the rock to earn our pay! I don't think our outfit is interested in some choice real estate," I remarked to Art.

"Yea. You can say that again. Gold is what they want, and if we finally find it, they will bury it again at Fort Knox. It don't make sense, not to me it don't. Us fighting the flies and eating bully-beef for that. However, good old Edmonton is coming closer. Let's go."

We swung north alongside of the rocky range, and then east, skirting a few large potholes on our way, we reached camp just as the sun was leaving us for the day. No mineral had rewarded our search, but we were content to call it a good day well spent, we had found what surely was one of the loveliest spots in the northern bush.

How it had come into existence puzzled us at first, but by and by on our future treks we found signs that seemed to give an answer. Faint vertical lines on the rocks facing the shoreline, high above the present water level clearly indicated that Hearne Lake had been much larger and deeper in times long past, and that the outlet had been where now our park was located.

On our map we had noticed that Hearne Lake was now emptying into Watta Lake, over a series of falls north-east from our camp. Occasionally, when the wind was right, we could hear a dull far-off roar, and we were anxious to have a look. For several days, though, we had to stay on the west side of the lake. Strong, cool winds were blowing, driving tattered, low-hanging clouds before them, and churning the waters of the lake into angry, white-crested rollers. Better to pull the canoe high up on the shore, and to depend on shanks' mare on our exploration trips inland from the western shoreline. It didn't take much imagination to see in the planed-off rocky ridges running north and south, separated by sand-filled depressions and swampy sloughs, the bottom of a much larger lake of the dim past, a mighty gouge in the face of the earth, filled by the waters of the receding ice at the end of a glacial period. It was interesting thus to speculate on what happened long, long ago, but from the prospectors' point of view the area covered was a disappointment. We could not even detect common minerals in the hard, solid rock.

Several nights and days were filled with the howling of the northern gales and the crashing of the waves, but at last came a morning when we awoke to a strange quietude. Light frost covered the ground, and when the sun climbed high into the clear sky, the leaves of birch and poplar came drifting down. "Soon we shall have to pull out," we said, as we prepared our breakfast, and like a confirmation, there came to our ears the sound of an approaching plane.

By the time we stepped outside, we saw a Junkers turning into our bay. After we caught hold of one wing, Wop and Mert Wales crawled out of the cockpit. "You fellows going to spend the winter here?" Wop called out.

"No, sir! We want to go home! Come in first and have some breakfast! The bacon is in the pan, and the coffee must be boiling, you can smell it from here."

"We are just making the rounds to find out who wants to go outside before freeze-up," Wop told us while we were having our repast. "Will a week from now be OK with you fellows?"

"It's all right with us, if you first take us back to Pensive, we shall have to cache our stuff for the winter, which will take us a few days."

"I don't understand why anybody wants to go outside; things surely don't look any too good there. There's a lot of war talk going on, nobody knows what's going to happen next," said Wop.

"Whatever happens, I've got to see my daughter finally, she will be four months old in a few days!"

"That's a good reason indeed! She is quite a girl!"

Our guests were in a hurry, and the day was still young when they roared away. The sun soon chased away the chill of the morning, calm and inviting was our lake, and so, at last, we set out to look for the falls, a warm southerly breeze helping to push us along.

Soon we found ourselves amidst a maze of small, wooded islands. Threading through the channels separating them we kept going east, expecting to see white water somewhere ahead of us as a guide towards the outlet. The water remained as smooth as a mirror. Suddenly our paddles found no resistance, but faster and faster we seemed to travel, an altogether eerie feeling!

Art, in the bow, gave me one look, swung his paddle over to the left side, and for all we were worth we paddled furiously toward the shore. Not a minute too soon. When our canoe grated against the rock we jumped ashore, our chests heaving. Hardly fifty feet farther on the waters were gliding over the brink!

"Holy mackinaw, that was a close one!"

"Yep, little Art almost didn't get to Edmonton!"

Through a dense tangle of undergrowth, alongside the roaring waters, we scrambled down to the level of Watta Lake. An awe-inspiring impression was our reward, deepened by the thought of our narrow escape. After the first smooth drop, like a glittering sheet of silver there was nothing but white water boiling over submerged ledges, with a few sharp rocks sticking up here and there, ending in a turbulent whirlpool, overhung by dancing rainbows; framed by the dark bush, it was an unforgettable sight.

"Don't think we would have made it," were Art's first words.

"Not a chance!" was all I could reply. "Not even in a barrel."

After a long while of looking on, we tore ourselves away, made our way back to the canoe, and a respectable distance away from the brink, crossed over to the north side. There we found a well-worn portage trail, and many old campsites close to the water's edge.

"We weren't the first ones here!" said Art.

"I wonder how many came as close as we did to taking a short cut?"

As usual, we had a good look around, but a few rusty tins and charred sticks didn't tell us much about the travelers of the past.

As we looked at the tumbling torrent pouring through the clean break in the rocky range, it wasn't difficult any more to visualize what had happened to Hearne Lake, and how our park had come into being where once the outlet had been at the south end. If one imagined a sudden break in the wall between Hearne and Watta Lake, and a sudden lowering of the lake level, the puzzle seemed to be solved. Art sometimes grinned when I thus speculated out loud, and at times he would say, "Why worry about that? I take things as they are, you always want to have an explanation."

"I'd just like to know the why and how, that's all, although I don't always expect to get an answer. Partly inherited and partly my training, I suppose," I answered thoughtfully.

"OK. Maybe you know the answer to this one. What made you come to this neck of the woods, and why do you stay here when you could be with your little family?"

"That's an easy one. I like the outdoor life; I don't like to do the same thing every day. I like a little adventure, and outside of that, I think we are taking part in pushing back one of the last frontiers. There are hardships, but a lot of compensations too. I really hope to make our home here in the North and find my life's work here. We have been successful. With a little luck, it should work out."

So, we talked while we lunched where the waters of Hearne Lake thunderously tumbled into Watta Lake.

"Maybe there will be a powerhouse here someday" mused my chum, "and you can have electric light on your homestead. Perhaps by that time though, they'll have figured out another way. In one of the magazines a guy wrote something about breaking up the atom. Said there was enough power in one lump of coal to drive an ocean-liner across the Atlantic. That'll be the day, I guess."

I replied, "The question is, 'Will we be any happier.' People always talk about progress, but are we getting anywhere? Years ago, my father gave me a book to read. *The study of Nature and Christianity* it was called or something like that. Anyway, the fellow who wrote it maintained mankind hadn't made much progress and gave some examples. He said, for instance, that the old Egyptians knew about as much about fine weaving as we do today. One princess had a shawl, big enough to wrap herself into completely, but fine enough so that she could pull it through her finger ring.'

"You had to tell me that. Boy, oh boy! You just give me that dame and her shawl, and I'll make progress and there won't be any lack of power either!"

"All right, let's go. It's time we get a move on! How about looking over the islands on our way back?"

"Suits me. Anything to get my mind off that blonde in the thin shawl."

"The Egyptians were black haired though, I believe!"

"Never mind, black or blonde. Don't make no difference to little Art. I ain't choosy. A little redhead would be OK though."

Well, we didn't even find any mineral on the islands. As Art said, "If this had been one of them South Sea islands, we would have had a chance to hear the rustling of a grass skirt maybe. It sure would have taken my mind off the gold. However, Edmonton is another day closer."

As the sun went to rest, we two weary travelers reached our camp. There was quite a chill in the air. A crackling fire was welcome and made a home out of the canvas walls.

CHAPTER XLIII
FLIGHT TO THE SOUTH

ONLY A FEW days later we were back at Pensive Lake, put our tent up once more, and got busy building a cache. All the stores and tools we wouldn't need for the remaining days we now piled onto the strong four-legged framework. We worked fast, because we wanted to spend some time yet on our glory hole, to pick some good nuggets for the folks at home, as a visible proof of our discovery. After the first warning of winter's nearness, the weather had turned very warm and pleasant again. Extended summer at Pensive Lake.

Early in the morning of the day fixed for our departure, our canoe, tent, stove, and the rest of our equipment was hoisted onto the cache, covered by a tarpaulin, and lashed down securely. Only our sleeping bags remained unrolled, mindful of the prospector's superstition that the plane won't arrive if you roll up your bedroll too early.

Around noon Mert Wales, with a Norseman, came for us. A last look around, and we climbed aboard. A few minutes later we reached Yellowknife. The town seemed very quiet. Many parties had pulled out before our turn had come, and now that we had left the bush, our friends and acquaintances in the settlement who had radios and read more papers than we did, immediately went to work to bring us up to date on the political situation, which was described as grim and foreboding.

Canadian Airways informed us that we would be leaving for the south the following morning, so we had a lot of time to make the rounds of the town. Everywhere we went we encountered an air of uncertainty and gloom, quite in contrast to the bustling atmosphere of the spring when everybody had been full of hopes and tall stories.

Fog was rolling off the big lake when at sunrise we assembled at the dock for our homeward flight. But, as the sun rose higher, the skies cleared to a fine day for flying. The water of Yellowknife Bay was glass smooth and we all crowded well forward in the cabin to facilitate the take-off. At long last the southward flight had begun. Tonight, if all went well, I would be with Lotti and our Marcia.

Already we were crossing the east end of Great Slave Lake. Soon now we would swing to the west and follow the Slave River. This time we were travelling in luxury, comfortably seated, not squeezed in on top of a lot of equipment and food as on most of our former flights.

Below, the now so familiar northern scenery, up ahead the motor droning its powerful song, by and by the eyes began to feel heavy, and then I must have dozed off. When I awoke, the sun was straight ahead on our course, and it seemed that we were going down. I looked out. Fort Smith was in sight. A crosswind but smooth landing on the swirling waters below the rapids.

We were just in time for lunch at the Hudson's Bay Hotel. The big dining room so crowded when we went north in the spring was almost empty, and the meal was served without delay. We don't waste any time. Soon we are down at the dock again.

Just then an RCMP plane pulled in. A very tall chap, a good friend from Edmonton stood on the float, ready to throw the mooring line. I hailed him, but he just waved back. Then I saw the reason. Two important-looking men in full uniform crawled out of the cabin, rear end first, something hard to accomplish in a dignified manner. The taller one of the two kept his head well down while he makes his way to the front of the pontoon, and so to terra firma.

"Ah, here comes the general now!" said one of our gang, as the shorter dignitary prepared to step ashore. He pulled himself up, and his tunic down, adjusted his cap to the one and only permissible angle, all very precise and snappy, then started to march. Too bad, he had such a good audience. An airplane strut was in the way. Cap and strut collided. Cap sailed into the drink. A roar went up from our gang, the disrespectful beggars!

But trust the RCMP. Our friend the corporal's paddle flashed and scooped up the important head gear. "Hurrah for the Mounties!" somebody hollered, and at the same time Mert's voice was heard, "Come on you guys, if you want to get to Edmonton!"

Next stop Fort Chipewyan. Two more passengers came aboard, and off we went again. To the east blinked big Lake Athabasca, our old stamping ground, and my thoughts wandered over familiar trails and campsites not far from its shores. For a while only, though, then they raced ahead of our southward drifting plane, trying to imagine the great moment when for the first time I would hug our first-born baby, hold the one I owe it to in my arms, and the yearning of six long, lonely months will have ceased.

So, I was dreaming, and it seemed the other passengers were busy with their own thoughts and dreams too. There wasn't much talk. Some had succumbed to the lullaby the motor is humming. I closed my eyes, hoping that sleep will shorten the journey. But the mind gave no rest and told the eyes to look for remembered places along the winding Athabasca to see how far we have gotten towards home. Then Fort McMurray, last stop before Cooking Lake and Edmonton. Wondering and pondering, a thought took shape and would not be driven out, however much I tried.

You are a damned fool, I told myself, entertaining thoughts like that. *Isn't this the day you were looking forward to with ever-growing impatience?*

True enough, chatted back my other self, *but your original plans were quite different. You wanted to make your home in the North! Once you two built a lovely cabin at Moore Lake, remember? You couldn't enjoy it for very long, true enough, but there are plenty of logs in the Yellowknife district to build another one on the shores of some enchanting lake. Didn't Lotti write in every letter, "When can we come to you?"* No use arguing against that. I knew then only too well that here lay the crux of the matter. There could only be one satisfying solution; to work with all my strength towards the goal of making the North our home.

Once upon a time we seemed so close to a realization of that. We had talked and planned a lot about our quarters at Beaverlodge Lake. At Pensive or Hearne Lake were good campsites too, but lately not much had been said about those big plans of ours. Pat had hardly shown up all summer, the attractions of Edmonton must have pushed his old dreams into the background. Were we just growing into a bunch of prospecting tramps?

Fort McMurray didn't hold us long. The day wasn't young anymore. Mert said the weather reports from Edmonton weren't any good, we'd better hurry and get there before it would close in.

Last lap of the long journey, and by and by the excitement grew. Nobody slept any more. Sizzling steaks at Johnsons, cool beer and blondes cropped up again.

Ronny got some expert advice from Art, "Now, look, this time for sure, we'll get you put. Big, husky guy like you. Cripes! Man, you are just a maiden's dream!"

Ronny just grinned and said, "First you have to get there."

"Another ten minutes and you can see the big town!" somebody answered, but that turned into a rash statement. The moment it was said, the skies turned dark, low-hanging clouds were scudding by, and then heaven and earth were blotted out. There was only grey, weaving, wafting nothingness around us. Mert stuck to his course, hoped for a break in the fog that might tell him where we are, he climbed a little, then came down again, the altimeter close to zero.

A roof suddenly loomed out of the pea soup, quickly he pulled his stick back. There was no more talk and banter. Everybody stared out of the windows, trying to see something where nothing showed. We must have been past Cooking Lake. We were flying in circles. But only one fellow got nervous, the same guy who had done the big talking all summer.

"Aw, shut up. Leave it to Mert to get us down," cracked Art.

Up front, Mert sat calmly at the controls, calmly turning his head from side to side, looking for a hole. Minute after minute dragged by, and it got darker and darker.

Finally, Mert turned around and said quietly, "Not much gas left, have to set her down soon. Hold on when I tell you!"

Hardly was this said when he pulled the plane around and into steep side-slipping dive. "Water below," someone called out with relief.

We felt the pontoons hitting the water, then a terrific jolt threw us all forward. The tail seemed to come up but settled back. We were stuck in the mud, that was sure, but where were we?

"Everybody OK?" Mert asked.

"Sure, we are all fine!"

To the left of the plane, some dim, dark outlines may have been land. A few of us started towards them, through just about a foot of water. It was land, all right, muddy, reed grown banks, and scraggly bush beyond. Not a house in sight. Shouts come from the plane, and from far off the sound of a kicker. Then, dim lights appeared.

We rushed back to the plane, where a boat arrived with two fellows in it. "How did you guys get here?" one of them asked. "Tell us first where we are!"

"South Cooking Lake! Didn't you know? They heard you at Cooking Lake for a long time. Phoned here too, they knew you were in trouble. We'd better call them now."

"How about you fellows taking some of our load ashore?" said Mert. "Once this crate is empty, I might be able to move it towards the shore."

Soon, our dunnage was ashore, and the plane moored safely, two cars and a station wagon arrived to rush us to Edmonton. That almost seemed the longest part of the journey, but at long last the longed-for moment has arrived and there was Lotti and the world seemed bright.

And then, Lotti took me by the hand, "Come and have a look at your daughter! She is asleep, but maybe she knows her daddy is coming, I have told her so often!"

There she lay, chubby and healthy, and my heart grew big with love and tenderness towards our little one and her Mummy as we stood there arm in arm, and I knew I'd come home. Little arms began to move, a little head with just a little bit of white-blonde fuzz on it turned, blue eyes opened wide, took one look at the strange, dark-skinned creature that stood close to Mummy, and then sweetheart cut loose with an awful howl of protest.

"Don't worry, Daddy. How do you like that? 'Daddy'? She'll get used to you!"

Before the evening was over, my little daughter appeared quite content to be held by her daddy, although in spite of a good scrubbing and shave, he still looked like a man who had spent his whole life in the bush.

CHAPTER XLIV
WINTER IN TOWN

AND SO, WE lived through a long winter of many uncertainties about the future, but rich in happiness and contentment. We watched with never-ending delight the growing of body and mind of our daughter, Marcia.

"She's the best little help against the blues!" we said when our whole Northern venture seemed in question. The big boss wouldn't commit himself concerning any future undertakings, and more and more often people were asking, "Will there be war?"

It was 1939 now. April came and May. Marcia had her first birthday, and still, we didn't know whether to get the bedroll out of storage and the old duffel bag packed. We only knew we had missed a lot of good prospecting weather by not getting north before break-up!

June came with great heat and the sultry and oppressive atmosphere was the perfect replica of the ever more threatening political situation. Finally, we were ready to give up hope, and look for some other way of making a living. At the end of June there came a call from Pat, "I've just had word from W.M. We are to do the assessment work on the Yellowknife claims. I want you to come to Calgary with me to see Howson about the details! We leave tomorrow morning early. O.K?"

"Brother, you took a load off my mind!"

"All right! Be seeing you."

The sun hadn't risen very far when we got going next morning. Pat surely got the best out of his Dodge that forenoon. The gas pedal down to the floorboard, we roared along in great style over the not too smooth gravel highway. "Only way to travel this washboard is to go about 65-70," growled Pat, who, although just as anxious as I to get things going again, didn't seem to look forward to the meeting with the moneyman in Calgary. To get that over with, and at the same time to get something off his chest gave two good reasons to bear down hard on the accelerator.

"Go ahead and step on it, don't mind me. I would like to know where we stand too. Don't get this contraption mixed up with old plane though. It ain't got wings," I said.

"If this goes on much longer, we'll have them. Damn it all, man, we have done good work, given them returns on their money, and now we practically have to go down on our knees so we can go back again to fight the flies, and risk our necks flying into those goddam potholes down North."

"You said it. But why don't we make ourselves independent from those fellows. After all, we know our way around the North by now, and if we are as lucky as we have been in the past, we should be able to make a go of it."

"That's the idea. But try and raise the money! The banks have more funds on hand than they know what to do with. Ask them for a loan though, and they want more collateral than the loan amounts to. However, this new set-up we are trying to arrange may give us a chance."

"What kind of a fellow is that man Howson we are going to see?"

"Well, he isn't exactly a friend of mine. As a matter of fact, he doesn't like me, then that's one of the reasons I took you along. But don't let that worry you."

By the time we reached Calgary, we had talked ourselves into a pretty warlike mood. It probably was a good thing that we arrived at noon and had to wait till the tycoon returned from his lunch. Thus, while we cooled our heels, we could cool our tempers too.

"Beer parlor is as good a place as any to kill the time," suggested Pat.

"Just lead the way. You know this burg!"

"No use being bashful about this. Ever hear of the Palliser? Maybe we are on our way up again!"

Quite a few cool ones and a precautionary call at a drugstore later, the fateful rendezvous took place. It turned out much more satisfactorily than we had thought, what with Pat even turning on his charm, and the important man beaming quite fatherly at us with big blue eyes. No, it wasn't bad, as a matter of fact, a pretty good arrangement was arrived at, but we were careful not to put on smug grins until after we left the august presence. The sun was smiling somewhat brighter too, it must have, it was extremely hot. We were dry and suddenly very hungry. We hurriedly corrected that.

"Let's buy some little things for the girls and the babies!" Pat then suggested. "We are in the money again. After that, let's get the hell home. There's work to be done!"

And so, after a little shopping spree and a short sentimental journey through Pat's old haunts, we wended our way homeward. If we didn't set up a new speed record on our way to Calgary, maybe we did on our way back, or was it that we felt so much better, that we talked so much about the work that lay ahead?

By the time we got home, our plans were ready. Within a few days the needed supplies were ordered. These included a gasoline driven portable rock drill, an expensive item, only purchased after tests, and a long weighing of the pros and cons. We were now financially more or less on our own, having taken the job of doing the assessment work on a contract basis. So, the motto ran: Watch the expenses.

We never had reason to regret the purchase of the Warsop drill, it saved us a lot of tiring and slow jackhammer wielding and speeded our work immensely.

CHAPTER XLV
NORTHWARD AGAIN

IT WAS PARTING time once more, but the relief from financial worry and idleness and the strong hope of an early reunion tempered the sorrow of having to leave my beloved ones behind. Once more the road into the future, our Northern future, looked clear.

Through the splendid summer day, we flew the familiar trail, touched down at the familiar stopping place. Fort McMurray was brooding in the heat, quieter perhaps than I had ever seen it, overlain by a smoky haze.

"Lots of bush fires farther north," somebody said down at the dock, and soon after taking off we noticed here and there below us big fiery half-circles eating their way through the bush and muskegs, and large stretches of land were obscured by drifting smoke. A series of dry summers had made a tinder box out of the northern bush. Sloughs, potholes and muskegs had dried up, and the level even of the big lakes was many feet below normal. At Fort Chipewyan, Fort Smith, and Fort Resolution the docks, once just a few feet above the water were now sticking grotesquely into the air practically useless for the airplanes. I was happy to notice that the atmosphere grew clearer as we approached Great Slave Lake. It would have been too bad if fires had been raging on or near our claims!

"I'll be along in a few days. Hire two or three fellows when you get to Yellowknife, if possible, from the old gang. Be all set to go to Pensive as soon as I join you," those were Pat's instructions. As we circled for a landing, I wondered whom I might find. The answer came quickly. One lonely figure was standing on the dock. When the float grated against the pilings, I made out our old powder monkey Fred.

"Hey there, Fred, you old sinner. What are you doing here?" I greeted him.

"Jeeskrist, Hermann, where are you coming from? Can you lend me five bucks?"

"Sure, I could, but I won't. Pat said, 'Hire the old so and so if he promises to stay sober till we get to camp'."

"Aw, come on. You don't know how dry I am. This is a hell of a dead place now!"

"Almost looks as if you were the only one alive here, Fred. What's the matter?"

"Buy me a drink and I'll tell you. When is Pat coming? I am going to get one out of him for sure."

"He'll be here in a few days. Give me a hand with my junk and tell me who of the old gang is around and looking for work. Then we'll see what we can do about the thirst."

"Saw Steve this morning, he ain't working steady. Has been cutting wood for some guy. Art has been hanging around here too."

"Couldn't be better. If you see them, tell them I'll be at Vic's. First I have to go and order grub from Mike's."

"I'll tag along if you advance me those five bucks afterwards."

I must confess I weakened to the persistent plea of the old reprobate, and thereafter didn't see him until two evenings later, when Pat roared in from the south. Fred must have gotten word to Steve and Art though, before he got too far away from reality. Both showed up the same afternoon for a grand reunion, but none of them knew where old Fred had gone to.

However, just when we had tied Pat's plane to the dock, old Fred joined our party, eyes bloodshot and generally worse for wear. With a wicked grin on his face, Pat stepped ashore.

"Holy cats! What's this? Constellation family reunion? These the guys you hired? Fred, you old booze-hound, you haven't drawn a sober breath since I last saw you!"

"I am on the wagon now, Pat, for sure."

"Well, there won't be any drinks in our camp. We got work to do. All set for Pensive? Let's get a move on." This was the stuff. Almost like when we first hit the North! In no time at all we had all our dunnage and grub at the dock.

Returning to Pensive was like going home, and shortly tents were blossoming again over last year's log walls. No camp was ever set up quicker by us. Our old tin stove came down from the cache, and while I got supper under way, Pat dragged our drill to the nearest rock outcrop, to give it another try. In an orgy of colour, the sun sank in the north-west when we finally sat down to bully-beef with tomatoes, canned peaches and the inevitable coffee and powdered milk. Mosquitos were humming outside the tent wall, two loons held an eerie conversation across the mirror-smooth lake, the stars came out by the thousands, the bush stood black against the luminous yellow-green horizon.

I had come home to the North.

After a long interval of involuntary idleness, we had put in some good honest work. Now we were tired, just plainly tired, from the physical effort. We would hit the hay soon, and we would sleep as we hadn't slept for a long time, deep and dreamless, without waking up in the middle of the night to be plagued by worrisome thoughts. The cobwebs had been blown from our brains; our bodies seemed to be suffused with a tingling feeling of well-being.

"This is the life! Wish I could make somebody see it!" Pat said as he zipped up his eiderdown. "Good night!"

"Good night!"

The next thing I knew, I glimpsed Pat stretching himself near the open tent flap. Was morning here already? "Come on!" Pat said. "High time to get going! There's work to be done!"

We ran to the lake, dashed some water into our faces, and while Pat mobilized the rest of the gang with some choice and powerful exhortations, I assumed a familiar role and got our stove going. A big potful of mush, stacks of flapjacks, bacon and eggs disappeared as quickly as I could put them on the table, accompanied by uncounted mugs of java. Food never tasted as good as on those crisp northern mornings after a good sleep with just a layer of canvas between us and the stars.

CHAPTER XLVI

ASSESSMENT WORK

OUR PLAN WAS to do the assessment work on the oldest claim-groups first. So, we set out in our two canoes loaded down with powder, fuses, the drill with spare bits, gas and oil, picks and shovels and grub. Having only one kicker, we lashed the two canoes together, and sort of zig-zagged across the lake, greeted by wild shrieks of our old friends the terns, power-diving within an inch of our scalps when we passed their nesting places on the numerous rocky islets.

In Steve we had hired the perfect packhorse for the drill. Nonchalantly he swung the contraption, weighing a good hundred pounds, onto one shoulder, grabbed the can with the gas and oil mixture in the other hand, and set off for the showing. We others divided the rest of the load amongst ourselves and followed the guy with the broad shoulders.

Soon the drill was pounding, and the steel of pick and shovel was ringing, clearing schisted, weathered rock away from the lead. Old Fred would squint at the rock, then give his orders as to where to put the holes and at what angle. With gratifying speed, the holes were punched in a row across the vein.

"OK now. You fellows better move on! I'm going to load and shoot!" Fred, who, in the meantime had measured and cut his fuses and crimped on the caps, told us. We took our drill and stuff and moved on. Only a few minutes later we heard "Fi-er-r-r, Fi-er-r-r!"

A dust cloud shot up, then: Bang, bang, bang, bang! Carefully spaced explosions shattered the silence of the bush. Rock fragments whined by our ears, as we crouched behind tree-trunks or boulders, listening for the rumble of rock torn loose, and of smaller pieces pelting down like hail.

"All over-r-r!" and we sprinted over to the neat trench. Fred's mastery of the powder business had cut across the veins, to see if any treasures have been unlocked. All day, with just a short interval for lunch, we kept at it, drilling, picking, shoveling, blasting, and sampling.

It was hot in the draw, there were plenty of mosquitos around, no doubt attracted especially by the streams of perspiration that poured off us, who were soft after such long inactivity. But we were feeling good, work was like a tonic. We didn't look at our watches, the lengthening shadows only told us that it was time to return to camp. When the sun goes down in those latitudes, it has made a long journey.

On the claims we were working now, no spectacular gold nuggets had ever been seen, only minute specks hardly discernible with the naked eye had been detected over a considerable width and length. Our trenching work showed the same picture, the lead was shot through with tiny particles of the yellow metal. This was encouraging, but we all were eagerly looking forward to the day when we would take our drill to the glory-hole on the Joy claims.

Our feet were dragging a little when we stepped ashore at the campsite, but what did we have a nice sandy beach for? A good soak, a hearty supper and we felt like tackling another job. In Yellowknife we had hunted up a few pieces of two by fours and some bolts with the rough idea of making brackets for our canoes to clamp on the kicker. A much better idea popped into my head, however, when I started the job. "What do you say, Pat, we bolt the two canoes together, alongside each other, far enough apart though, to let the kicker work between them at the stern end? That will give us a steadier craft, and I feel sure it won't zig-zag anymore."

"Sounds OK. Let's try it that way."

With the last bolt tightened, Pat clamped the kicker onto the stern most crosspiece, jerked it to life and off we went. With the throttle wide open we skimmed over the glassy water in great style. Only in the turns we lost most of our speed but moving the crossbar and kicker still farther towards the stern improved the performance. As Pat put it, "Now she turns on a dime! A little more power and this thing will take off."

Once more the sun set in a riot of color, the loons took up their sad discourse, a duck family floated by with a few muted remarks. Just time to write a few lines to our families, a last smoke, a last look over the lake and so to bed.

"Sure feels good to be honestly tired," I said to Pat, "or do you want me to sing you a lullaby?"

"Not that, surely. Good thing I don't need it. Cripes, I am tired! First time in my life I worked like this. Something can be said for it. Keeps you from worrying."

"Yea. Only a few days away from the city, no newspapers, no radio, and one almost comes to think that all is just fine in this world. Is it though?"

"Who knows? Maybe we are just kidding ourselves! In any case it's time to catch some shut-eye."

A few days of that kind of work, and we were in good trim again. Our evening conversations grew longer. There was much to be discussed, that had been left unsaid in the past, things which had puzzled me, and which to my regret had clouded the friendship as it once had existed between Pat and myself. It was high time to clear the atmosphere.

I had to ask many questions to find out what changed Pat's whole outlook and especially why he, who often proudly said he never let an old friend down, had listened to the newcomers who, to his face, acted like friends but behind his back had nothing good to say about him, and on more than one occasion had betrayed his confidence to further their own aims. Harsh truth had to be spoken at last, and we went at it hot and heavy. I like to think that it was a measure of what was left of our friendship that we didn't end up by completely falling out, but instead came out of it with a better understanding of each other.

"Damn it all, man, you are persistent! Why in hell do you want to dig up these old affairs?" Pat exclaimed more than once in answer to my starting with "I'd like to know…"

And so, we went over all those disturbing events of which I had the first inkling when Bob Leonard told me at Goldfields, "I am not going back to that camp."

"Yes. I'd like to know why Bob left, who trudged by himself through that bitter cold night to Goldfields to find out what had happened to you. There was a guy! He was a friend! I'd like to know too why Moore Lake was handed on a platter to those false friends of yours, although it meant breaking your given word which we had accepted unquestioningly! Why did you let your real friends down last summer by not spending more time in the bush, where you and the plane were needed? All summer we had to listen to the squawks and denunciations by the other guy when you were away. "Little bastard" was the mildest name pinned on you, but when you showed up for a few days, someone fawned and flattered, and you swallowed it. That's enough to make anybody sick and tired, especially as we knew that someone was only too glad that you weren't around more often, so that he could make side-deals with some of his friends who were staking around us before we were through, I want you to know."

"Holy cats! Aren't you through yet?"

"It should be enough! 'No damned yes-men in this outfit' isn't that what you used to say? That's why I told you. I want to make sure the same thing doesn't happen again. We are making a new start, for the love of Mike, let it be a clean one!"

Deep into the night we thus wrestled with the past in order to blaze a trail to a better future. When we finally dug out our last bottle of beer, I felt quite confident that things would work out right, and I believe Pat felt better too. "Well, here's to a better understanding!"

Work progressed well, we moved a lot of rock, and praised our little drill which saved us so much work in the blistering heat that scorched the North. One cloudless day followed the other, with a red-tinged sun that beat down mercilessly through a smoky haze. Lazily shifting light breezes sometimes brought the acrid smell of burning bushland to our nostrils.

"Fires must be close by," we said, and spent a day widening the fireguard we had cut at the base of the peninsula where our camp was located. The bush was bone dry, so were the sloughs and potholes, and the level of the lake continued to fall perceptibly. "Fires all over the place," reported Tom Mahon, the pilot who dropped in every few days to look after our needs and bring our mail.

At last came the eagerly awaited day when we could move to the high-grade show. We were not disappointed. The wider and deeper our test pit grew, the more gold, in bigger nuggets showed up after every blast. Drill, blast and muck out, again and again it was repeated, and eagerly we rushed back into the pit before the dust had settled after the shots to look for the gleaming yellow stuff. We had the gold fever all right.

It could have gone on for a long time without losing its attraction, but some other claim-groups needed our attention. Assessment work had to be done and the aluminum claim-tags had to be nailed to the claim posts. Pat came out with it one evening when we returned to camp, "When do you figure on going to Loon Lake?"

"I wish we could forget about that. It's just a bloody wild goose chase, to be sure!" I answered. I had no confidence in that set-up at all. During the last winter, Woods who had just received his private flying license had talked the big boss into flying to that place to stake some claims on the strength of a mysterious rich sample. He had persuaded Dan to go with him. They got there all right, after what Dan later on described as a nerve-wracking flight just above the treetops, staked claims and came out again. The whole thing sounded pretty crazy to me when I had first heard about it, and I wished I had had a chance to talk about it with Dan who, unfortunately, wasn't around when we left Edmonton.

Being without a job, it hadn't been very difficult for Woods to talk Dan into joining him in another crazy venture. Somewhere in B.C. there was an old placer mine. Some fellows in Edmonton thought they had a claim to it. It was to be a real bonanza, and for all I knew, it might have turned out to be one. Maybe they just picked the wrong guy to hold the water back that constantly flooded the diggings. In any case, it ended pretty poorly. However, this I learned only much later when Dan came back to work with us, and also told the story of the Loon Lake claims. By then, I had drawn my own conclusions.

The next time Tom Mahon flew into our camp, we arranged to have him fly our party into Loon Lake as soon as possible. I had counted on Pat going with us, of course, and was quite disappointed when he told me at the last minute he wanted to stay in camp, maybe go to Yellowknife for a day. I tried hard to talk him out of it, but he wouldn't budge.

"Why do you want me to go with you? Four men should be enough for the job, and that will be a good load for the plane anyway," he said.

"The reason I want you to come is that you might see for yourself that no gold came from there. Art and I were at Loon Lake last year, don't forget. We saw the samples they took," I replied.

"What do mean?"

"We'll talk about it some more when we get back from there. Here comes Tom now. I'll bet you anything we won't even find mineral, much less gold. You won't change your mind and come along?"

The other boys were almost through shoving our gear into the plane. No use hanging around any longer. "Well, so long then, Pat. Don't be talking to yourself by the time we get back!"

"So long, Squarehead. I'll have one on you at Vic's!"

As it turned out we had to make two flights. The water was glass smooth, but Tom couldn't get his crate into the air until we had lightened the load. Old Fred and I went first. We were hoping Tom might be able to take us right into a small squarish lake, fairly well in the center of the claim-group, but after circling around a few times, he decided otherwise and landed on the main lake near the mouth of long, narrow arm of water running east along the north border of the claim-group.

Tom was off as soon as we had untied our canoe from the pontoon struts and unloaded our other gear. It was a long wait till the second load arrived. Steve and Art were in high spirits.

"Made a little side trip to Yellowknife, had us a few beers. Pat went out with us. He bought a few bottles for you guys," Art reported.

"OK. We'll have it tonight. Let's get going to get our camp set up. The day is half gone."

So, we paddled down the channel until we arrived at a favorable spot which we figured was about half-way across the top of the famous claim-group.

"Jeeskrist, but this is gosh awful looking country," Art remarked, and that's about all anybody could say about our new surroundings. It was a barren waste of rock, all that was left after a forest fire, not so long ago. Near the water's edge a few trees were standing but farther inland only the first timid beginnings of a new vegetation.

"If there's any mineral around, we won't be able to overlook it, that's sure. But how in hell did those guys stake the ground. There is hardly a tree big enough from which to cut posts for miles around. We'll have a sweet time finding their lines, I'll bet you. However, let's get our tents up first. I feel I need some beer too." I had a premonition that a few unpleasant days were coming up.

A few rounds of bone-dry charred logs served as basis for our cook tent. At one end I fixed my bunk. The remaining space was taken up by our grub boxes and the old tin stove. The other tent was set up about twenty feet away. Evening was falling when everything was arranged. We had a sketchy supper, and then went for the beer. There we were four fellows right in the middle of nowhere. Steve softly blew a few sentimental notes on his mouth organ, old Fred was telling tall stories about days gone by in San Francisco, Butte, Spokane, and many other places he had drifted through as a miner and powder monkey, working for shorter or longer spells, then blowing the hard-earned stake in one grand and glorious spree.

Good resolutions to make the case of beer last for several days soon were forgotten. When the last bottle was gone, we were ready to retire. Zipping open my bedroll, I made sure that the bottle of Hudson's Bay Rum I carried for emergencies was safely tucked away at the foot end. That's the last thing I clearly remembered the next morning, and the fact that I once woke up to hear loud singing from the other tent, marveling what a few bottles of beer could do.

It was stifling hot in my tent when I came to with a start and crawled out of the sticky sleeping bag, almost dizzy with the heat. I threw open the tent flaps. The sun was high, beating down mercilessly through a smoky haze. A real scorcher was underway. Nine o'clock in the morning! We had slept through the best working hours. Not used to celebrating anymore!

Not a sound from the other tent. I put on some coffee, then I called for my chums. No answer. Damn funny business. I walked over, pulled the tent flap open. Wow! The heat was even worse than it had been in my tent, and there was quite an aroma to it.

"Hey, you guys! Wake up! We've got work to do." Steve turned onto his back, started snoring. Art just grunted, shifted a little, and was still again.

Old Fred opened one bloodshot eye, closed it as if the light was hurting, muttered something, and he too went into the snoring act.

"Hey, Fellows! Time to get up!" I hollered. No results. I had to pull and shake for all I was worth before I finally got some life into them, and they weren't any too happy about it. Only a few bottles of beer, and one hell of a hangover. It sure was a puzzle. By and by I had my three chums sitting up, at least, holding their heads.

"Got any beer left? Cripes, I am dry!" Was the first comment from old Fred.

"Coffee will do you more good! Come on, fellows, it's high time we get going!"

"Maybe it was the heat that got them." I pondered, as I went back to the cook tent. Then it dawned on me, the bottle of rum! I reached into the foot end of my bedroll; the bottle was gone!

Three men, three bottles of beer each, 40 ounces of Hudson's Bay Rum on top of that, and then the heat. I wasn't puzzled anymore! One by one my three revelers, gingerly, as if every step was jarring their brains passed the cook tent down to the lake to cool their heads. Appetites weren't very big that morning, only coffee was in demand.

"You fellows had yourself quite a party, it seems," I teased them a little. "How was the Rum?"

"Rum?! We never saw any!" came the chorus.

"Don't tell me you got that tight on a few bottles of beer. Might as well come clean, how did you get it out?"

"OK. We got it all right. Nothing to it. You slept like a log, and your bedroll wasn't even zipped up. It was good stuff! How I wish I could have a shot now!"

"Better have some more coffee. We've got a long, hard day ahead of us. Here is the set-up. According to this fancy little map the northernmost claim-line runs somewhere along this channel. Look! Here is a mark. Supposed to be the location of the mineralized lead, where, if we can find it, we want to do our assessment work and take samples. One of you, Art, or Steve, will leave to start putting the tags on, or we'll never get through with our work. We'll all help to locate the first claim post, that should make it easier."

Art volunteered for the job of tagging. Only a short distance from camp we actually found a claim post from which he could start off on his lonely mission, while we took our tools and supplies for the day in the canoe as far east as we could go along the channel. We figured the overland trip couldn't be very long, if the mineral actually was where the sketch showed it.

We were optimists. Soon, however, our outlook changed. Through a nightmarish welter of burnt and fallen trees, we battled our way south. Encumbered by our gear, we staggered, slipped, and fell. Sweat poured from us and attracted clouds of black flies to add to our troubles. At last, we got onto more open ground, where a breeze brought some relief. We dumped our gear, rolled a smoke, and began a careful search for the mineralized lead or outcrop. Hour after hour we zig-zagged over the country, neither seeing the slightest sign of mineral nor of any prospecting activity. Nowhere did we see that someone had broken rock, but what was still more disturbing, we didn't see any blazes or posts.

Said Steve, "Art must have a hell of a time getting rid of his tags if even the three of us cannot find a claim post."

"I wish Pat were here to see this mess! He couldn't see why I hated this lunatic proposition! What are we going to do next?"

Old Fred, pretty well tuckered out, came right back, "One thing is sure. I am not going to do much more of this wild hunting around. Let the guys who staked these claims look after them."

"OK Fred. I don't blame you. I am fed up too. You make your way back towards the canoe but take the powder and caps along. If you see any good cracks where you can move some rock, load 'em up and shoot. Might as well put up a good show. Steve and I will do a little more looking around before we join you."

After a little snack we parted. Old Fred made a beeline northward, Steve and I continued our fruitless search. Off and on the thunder of Fred's operations shattered the stillness. At least we got some assessment work done!

Once more we circled far to the east. Again, the result was nil. We had had enough. So, we picked up our gear, and made off in the direction of the last shots, taking good care to avoid the tangled deadfalls through which we had lugged ourselves and our loads in the morning. Before we found Fred though, we were hailed by Art from a rocky knoll not far off. Our chum did not look happy.

"What did you find, Art?"

"Find? Well, that's just the trouble, you can't find a damn thing. Worst goddam staking job I've ever come across. No lines, no nothing! I've still got most of the tags! Poor little Art had a hell of a day!"

"So had we. It turned out worse than I feared."

"What's the shooting about?"

"That's old Fred doing assessment work! He seems to be having a good time. Let's go and find him."

Soon we spotted the old boy. He was busy picking over the rock he had blasted from a small shear zone he had run into. It even showed some mineral, but nothing to get excited about. In any case, it was far from the cross on the map, but just to make sure I asked him, "Seen any signs of prospecting here before you put your shots?"

"Hell no, man. Bet you dollars to doughnuts I was the first man to set foot here. Only a prospector is crazy enough to wander into this bloody corner. The Indigenous people are smarter."

"Let's take some samples and head for the canoe."

"That's the first intelligent thing that's been said today."

CHAPTER XLVII

A LOT OF SMOKE

IN THE AFTERNOON, a light breeze from the west, that smelled strongly of smoke, sprung up. As we quietly glided towards camp, the smell of something smoldering grew stronger. "The fire must be damn close," someone remarked, then after a while someone else said, "Smells strange. Not like a bush fire!"

"It's our camp!" we all must have cried together. Our paddles flew as we raced the last few hundred yards. Then for a moment our hopes rose, shining white through the haze there stood a tent. Where was the cook-tent though? A few minutes later we stood in front of the smoldering ruins. Nothing was left, even the logs were just glowing embers. A sad bunch of fellows poked around in the debris.

Only a few scorched spaghetti strands, a little flour and a small piece of bacon were salvaged.

My bedroll and spare clothes were burnt to a frazzle, but, saddest of all, my reflex camera was just a lump of half-melted metal. "Can't even take a picture of this mess. We'll have a lean time before Tom gets back two days from now!"

"Never mind the picture. We won't forget this in a hurry. We'll be hungrier than them starving Armenians." Art came back, adding after a thoughtful pause, "I am only glad we had that rum last night!"

All we could permit ourselves for supper were a few slivers of fried bacon. Steve, a gargantuan eater normally, looked pretty glum and wistfully commented, "Bet you even bully-beef will taste good when we get back to camp."

"Aw, nuts! Why did you ever leave the farm, Steve? All you can ever think of is eat, eat, eat. The farm is the place to get it, and you bugger off into this blooming wilderness! Serves you right." Art was rubbing it in. Pretty soon, however, we were all spinning the same stories, the hungrier we got the more we talked about the fancy feeds we were going to treat ourselves to at the first opportunity.

However, things could have been considerably worse. That our second tent had been saved was a miracle. Its canvas bore innumerable brownish marks where flying sparks had landed and died out. It doesn't take much to set a tent on fire. This one had a truly miraculous escape, and thankful for small mercies, we shared the three bedrolls for a fairly good night's sleep. Preparing breakfast didn't take long the following morning. Again, we had slivers of bacon, and fried in the grease a little paste of flour and water. Then Steve, Art and I set out to see what we could do about tagging the claims. Old Fred we left behind to move some more rock on the cliffs near the water's edge.

The air shimmered with the rising heat as we strained our eyes to find the blazes of claim-lines to lead us to the claim-posts. It was bad, and worse even than Art had described it, but we struggled on, desperately determined to finish the job. We were tired, hungry, and mad as evening drew near, and yet we still had half of the tags left. Another day of it to come, and hardly anything left to eat!

We had heard a plane off and on during the day, and when it seemed to come nearer, our hopes rose that we might be able to flag it down. It never came in sight. After supper, a few strands of spaghetti boiled in water, with a few cubes of bacon, we were still so hungry that we came to a desperate decision. "Maybe there are fish in this channel; we could use dynamite."

We thought we had every excuse to try it. Old Fred prepared some shots, and we canoed towards the main lake. Overboard went the first shot and we beat it. A muffled roar, a short sharp vibration, and back we went to scoop in the fish. A few gas bubbles were still rising, but there was nothing floating belly up. We picked another spot. When the fuse was sparking away, Art heaved it overboard. Something whitish gleamed on the surface of the disturbed water. With all our might we made for it. But not fast enough! We were almost up to it when a seagull swooped down, scooped up our prospective supper, it wasn't a small one either and sailed off. "Son of a gun! Look at that. He's got a payload." There was nothing else.

"Let's try again!" We turned to Fred.

"I am sorry, fellows! Those were the last two caps I had."

The hollow rumble of our stomachs expressed our feelings, and sadly we went back to camp, to chew on the last of the bacon, including the rind. We had some trouble going to sleep on our empty stomachs.

"Tomorrow Tom will be here!" we said it over and over again, but tomorrow seemed a long, long way off. We had been living so well before, that's probably why we felt the lack of substantial chow keenly. Food had become an obsession, and the less we had the more we talked about it, calling up enticing visions of steaks smothered in mushrooms, Art's specialty, or half dozen new laid eggs with fried ham, pancakes with home-made jam, etc., Steve's version of a man's breakfast.

Fred wasn't particular, as long as he could come close to some honest food once more. I remember I had beautiful visions of a rare roast beef but assured the other fellows that I would settle any time for a choice cut from a horse, as the Norwegians know how to prepare it.

One more day we trudged through the bush, hunting claim-posts. In the evening, we still had a few tags left. But we didn't care much anymore. There was absolutely nothing left to eat. We smoked incessantly; it helped a little. Very early we crept under the sleeping bags. At dawn we broke camp and started to move our rock samples and our gear, what little was left of camp to the place where we expected Tom to land his plane. Of course, we were far too early. It was almost noon before we finally heard the most welcome sound in the North, the drone of a long-awaited plane. It was Tom Mahon all right, skimming toward us in his old Fairchild.

He leaned out of the cockpit when he was within hailing distance and shouted, "Get you tonight, fellows! Have to make an urgent trip to the Dome camp! Somebody got hurt."

He was ready to pull away, but our frantic hollering and waving made him stop. Art and I jumped into the canoe and paddled to the plane for dear life. "You got anything to eat on your plane, Tom?"

"To eat? What's the matter with you guys? You bushed? You took a planeload with you!"

"Sure, sure! But we haven't eaten for three days. Our camp burned down!"

"Holy mackinaw! You really mean it. Just wait a minute!" With that Tom crawled around into the fuselage.

"Hey there! Catch!" he shouted and flying towards us came bully-beef cans, hardtack, sardines, tea and whatever else made this emergency ration.

"Will that hold you till tonight, fellows?"

"Sure, we are all right now. Thanks a lot!"

"OK See you later!" and off went the good Samaritan.

Our two chums were right there to help pull the canoe with its precious load (food, not people) ashore. We just opened the cans, and dug right in, and as Art said, "Good old bully-beef! Yessir! We live again!"

A lively breeze sprang up in the afternoon, and Tom had no trouble getting his well-loaded plane into the air, when he finally came to take us home. No rose-grown cottage could have looked more attractive to us that evening than the three white dots on that sandpit jutting into Pensive Lake, our tents at the base camp.

Pat was there to welcome us. He sort of squinted at us; the map of Ireland screwed up in quizzical expression.

"Holy cats!" he said, "what happened to you guys? You look like something the cat wouldn't drag in. Why'nt you shave once in a while?'"

"Here he goes again, boys! How I wish you had been with us, brother!"

"Come on, tell! The stuff any good over there?"

"There ain't any stuff! And our camp burnt down. Give us a smoke first. I'll fix something decent to eat for us. After that we'll tell you the whole story. If you have a drink cached away somewhere, better dig it out!"

"First you guys burn up half of our equipment because you got tight on a few bottles of beer, and then you expect me to buy you a drink! Nuts to that!"

"Take it easy now. This is a time when even an Irishman should be careful. We are still pretty mad. After we have eaten, we may feel better."

"As bad as that, eh? God bless my teeth. You guys aren't going to put all that away?" Pat's glance had wandered to the mound of food I had piled on the table. "We'll never make any money with you guys around."

That didn't worry us a bit as we silently worked our way through to the other side. It must have been impressive. For once Pat seemed speechless, but when Steve polished it off with two cans of peaches, he shook his head, "I wouldn't have believed it if I hadn't seen it! Pretty good hay, eh Steve? Why'nt you have a few more peaches?"

"I might at that, after a while, before going to bed," drawled Steve. "Any more java?"

"I give up! Well, let's have the dope on your work down there!"

He got our report, looked at the samples, which weren't very impressive, and at last he came to the inevitable conclusion, "The so and so sure put one over on us!"

"You said it!"

Pat went out, but in a minute, he was back, bottle in the right, tumbler in the left hand, crossing them he stretched them out, "Here you are. Help yourself!"

Our outlook was better, much better. A big moon was pouring silver over the quiet waters of Pensive Lake, far off a loon was crying, an owl flew by on silken wings as we smoked contentedly in front of our tents. Replete with good food and a cheering drink, life once more looked good.

Pat was unusually quiet, I noticed, but his pensiveness didn't give me any warning of what was to come. When at last we went into our tent to retire, he suddenly said, "I hate to tell you this, but I am going south tomorrow. There are reasons. Here, read this," he held out a letter to me.

"I don't want to read your private letters. If you say you have to go, that's good enough for me. But I am sorry to see you go. I was beginning to enjoy once more our being together in the bush, we seemed to have recaptured the spirit of the early days. You seemed quite happy too."

"I was until this damned letter arrived. I'll try to get back as soon as I can make it. By the way, things look worse than ever politically. People in Yellowknife all talk of war. However, that's nothing new, let's hope it won't come off. We have had these scares before. In any case, try to get all the remaining claims tagged, and do some assessment work on the Sal group. Better tell Tom tomorrow morning when you want to go there."

"OK I'll take care of things. But try hard to get back soon. I might miss you. Nobody to fight, let's say, verbal battles, with."

"If not sooner, I'll come when the big boss comes out, which should be soon. I have been thinking we might propose to him to work the high-grade on shares, after the assessment work is done, in case they don't decide on drilling."

"Sounds all right. Vic's gang is doing it. They make it pay. Our show is better, and a lot handier. We'll take a gander over there one of these days and look at their set-up."

"Well, good night then."

"Good night! I hope you won't find things as bad as they sound in that letter."

Our camp was astir early next morning. Pat was packing his dunnage bag, while I prepared morning chow. Art sharpened some drills while Steve and Fred bolted our two canoes together. Tom Mahon arrived just as I was banging away at our steel triangle. "Come and get it!"

It didn't take much coaxing to make Tom join us. "Smells real good here," he remarked. "Heck, you fellows are living well! Methinks I should turn prospector too, instead of hopping from one crazy little pothole to the other, hoping I won't wreck my crate on some damned reef. Yep, it's real nice and peaceful! Any more flapjacks left?"

When the Fairchild had roared away, we headed for our high-grade show. There wasn't much point in digging around in our glory hole, but as Art said, "Let's see some gold, and restore our faith in the North!"

So, we drilled, blasted, hammered, picked and altogether had ourselves a grand time, "Look at this one! How's this for a nugget?" "Just wait till little Art lugs this stuff into Birks to have himself a tiepin and a ring made!"

For a few days we indulged in working our high-grade hole, making a separate pile of all the rock chunks which showed free gold. "When we have a good pile, we'll crack it all some more, no use to pack a lot of rock," we said.

So far, the weather had been extremely hot and dry, with only light breezes smelling of burning bushland blowing across the wilderness. One afternoon though, we had just crawled out of our hole, where the heat was stifling, to rest from our labors and have us a smoke, when suddenly there came a strange sound of rushing air to our ears. Dry leaves, moss, dust swirled up, and before we quite knew what was going on, a funnel-shaped apparition raced towards us from the south-west and passed not a hundred feet from where we sat.

"What do you know? A twister!" somebody said, and then "look what's coming up! Let's beat it."

We had hardly reached our canoe when blue-black towering clouds blotted out the sun. In a matter of minutes fierce gusts of wind made a raging sea out of our peaceful, shallow lake. We just made it. Once in our tent, all hell broke loose outside. Blinding flashes of lightning stabbed earthward all around us, and the thunder made the ground vibrate. Wildly the veering gusts whipped the canvas of our tent, cracking it like pistol shots. At last, the rains came. First, big wind-driven drops, then a steady downpour.

We retired early after our supper. The coolness was just right for sleeping, and anyway, there wasn't anything else to do. All night it poured. Whenever I woke up, I heard it drumming on the taut canvas. When morning came, the skies cleared. The air smelled washed, but soon, too soon, it turned into a steaming hothouse atmosphere.

When we got up to our workings, we found our high-grade pit a quarter full of water. We always wanted water to wash the dust off the walls. Now we had ample, and the labor of bailing it out was richly rewarded by seeing the gold show up all over. Where it looked richest, we put our next shots, and so it went; drill, shoot and pick, until another day was gone.

It was a terribly fascinating business to blast where we knew the gold was, but the deeper we got into the solid rock, the more ineffective became our shots. After all, we could drill only short holes with our equipment. So, we finally decided to move along the line of strike to make headway with our assessment work.

One mineralized lead, which we took for the extension of our find, ran alongside a pronounced break in the formation, and here a few well-placed shots moved tons of rock in a hurry. We must have moved enough rock to hold the claims for ten years.

Several days after the agreed time, an airplane finally showed up one morning. Instead of making one turn into the wind and then come in for a landing, this one kept circling for a while before finally sitting down quite a way from camp. We were puzzled. The plane looked like the old familiar Fairchild all right, but Tom Mahon's approach was usually a little more brisk.

Slowly the plane came taxiing towards camp. An unfamiliar face peered out of the cockpit. We frantically motioned towards our dock, but it seemed the fellow didn't trust us. We had to go out in our canoe and tell him it was safe. He was very polite and correct, and it probably was a good thing he didn't hear any of the powerful remarks that passed between us roughnecks as we paddled over to him. It might have hurt his feelings. He was a sort of scholarly-looking gentleman, and quite a contrast to our other bush-pilot friends. We told him where we wanted to go, but he was quite positive four men and equipment would be too much for one load.

Art said he would stay behind and hold the fort.

As it turned out, our gentleman pilot had quite a time to get the three of us into the air. Maybe we were super critical, but there was no doubt we taxied more than we flew to our destination. Eventually we got there though, put up our tent, went to work tagging claims, and doing some blasting on a big quartz outcrop which was supposed to have been the reason for staking those claims in the first place. There was an abundance of quartz, but of a granular, milky white variety, with only the odd speck of hematite in it.

We were through with our work the following noon, but the plane didn't show up at the appointed time, which gave us a chance to do a little prospecting for a change. The results were negative, but at least we learned that we could eliminate that particular corner from our further speculations. After a sketchy supper, as we hadn't brought very much grub with us, we sat near the shore, and smoked, and talked, while the lowering sun set the skies aflame, and painted the scene with magic color. Old Fred was unusually talkative, told us about his early days in Switzerland, "Alpenglow" he said they called it over there, when the tips of the mountains still seemed afire, long after the light of day had faded from the valleys.

For the first time Fred told us the whole story of his life. We wandered through France with him, then to England, from there to the States, lived through the San Francisco

Earthquake, and trudged through the mud of Flanders during the first World War. Then he took us through practically all the well-known Canadian mines, from Sydney, N.S. to Trail, B.C., and that linked us up with the time when we had first gone into the North.

One thing I still wanted to know though. Often the boys had mentioned Fred's singing. I had never heard him, and I must admit I was somewhat skeptical. With the old fellow in such a peaceful mood, maybe I would have a chance to find out. So, at long last I said to him, "Fred I've heard the others say that you sing very well, how about singing for us now?"

"Aw, hell, I can't sing any more. I've hardly a tooth left! When I was young…"

"Well, then the other guys must have been pulling my leg, when they told me that story!"

Fred just looked at me, and then looking out over the lake, softly, very softly, he began to sing, Toselli's *Serenade*, so often heard before, but never as I heard it that night. I had to keep on looking at our old friend, as stronger and stronger grew his song, it seemed so incredible. There was all the passion and the yearning ringing out in a crystal-clear voice. Someone might tell me it couldn't be, that I must have been dreaming, but I would swear I heard Fred sing beautifully that night at Pensive Lake, and never was I more deeply stirred by any singing. To remember, I just closed my eyes, and I travelled back once more to the spot by the quiet waters of a northern lake, and I heard again my old chum pouring out his soul in immortal melodies.

My wonderment grew when Fred continued with excerpts from famous operas, but it reached its peak when he said, "Now I will sing from *Tosca* for you, as I remember Caruso singing it." He pulled all stops, carried away by the fire and emotion of his last tragic song. Who knew what memories were stirring in his soul? For me too, memories of days long gone came alive. Somebody would rave about a singer's performance at the opera, and my father would say, "A good voice, maybe, but you should have heard Caruso!"

Alas, I had heard the great Italian only on records, but that evening in Fred's singing he came alive. The song ended. Without another word, Fred got up and went into the tent, leaving us wondering and deeply moved.

It wasn't only the coolness of the night that made us shiver, I am sure, as Steve and I crawled silently into our tent, where Fred had already disappeared into his bedroll, alone with his memories of better days.

At noon, the following day, having spent the forenoon futilely waiting for the plane, I suggested to my chums that we hike it home to our camp. "Come on, let's go! It's not far and it's easy going. I was over the same ground last year."

I tried to make it sound quite attractive, but only reluctantly the two other fellows agreed to go as far as Ingraham's camp. Our belongings were quickly piled up, and the tent draped over them, weighed down with some rocks. With a little grub in a packsack, and armed with prospectors' picks, we set out along the lakeshore. It took us about an hour to get to Vic's mine. The set-up had been quite a bit improved, compared to the previous year, and our host, who welcomed us in true northern fashion, told us that it was definitely a paying proposition to work the high-grade.

Something had gone wrong with the machinery though, that day, but someone said, "just stick around 'til tomorrow, and we'll have her perking again."

Fred and Steve were quite agreeable to that suggestion, but I was determined to go on. "If you insist, you can go with us, as far as our workings," Vic's brother said, "we are going to bring some ore down, it's on your way." Refreshed by some good strong coffee which our neighbors insisted on brewing first, we hit the trail. Not very far inland we came to a small lake where a canoe was pulled ashore.

"Your showing is on the other side?" I asked.

"No, there's a short portage into another lake, where we also have a canoe. Smart arrangement, eh? Saves us a lot of packing. The rock is heavy."

A short distance beyond the far end of the second lake, we came to the place where the gold bearing ore had been found alongside a pronounced shear zone. While the others got busy bagging the high-grade, I had a good look at the formation which seemed to be very similar to our own find. Then we parted, and keeping to a high, level ridge, from where I had an occasional glimpse of Pensive Lake, I headed in the general direction of our camp.

The going was good, and the sun was still high in the sky when I arrived at the mouth of a long, narrow bay. On the far side was our camp. Trying to save myself some further walking, I hollered and whistled for all I was worth. Soon Art's answering shout came across the water, and then I saw him barging through the shrub willows and reeds. "Wait, I'll get the canoe!" he shouted when he saw me. A few minutes later I heard the kicker come to life, and our double-breasted craft came around the point.

Art was grinning and happy. "Boy, am I glad you came! Three days all alone is a hell of a long time! What happened? Where are the other guys?"

My story was soon told. "So, they wouldn't hike it with you? The lazy so-and-sos," Art commented. "Well, never mind. I am damn glad you at least came. You know what? I was beginning to talk to myself. Now I know how guys go bushed. It's easy!"

Two days later the other half of our quartet finally came home. They had made a detour to Yellowknife, which they reported as "dead," as lots of fellows were pulling out. Altogether, not a rosy picture. But that did not deter us from keeping on with our work. Every morning we went up to the workings, to drill and blast and muck some more, and it was easy to forget the threatening world situation. We worried more about the fact that Pat didn't show up, nor even send any information concerning future work. Another idle winter in Edmonton was a grim prospect. We definitely couldn't see any reason why there shouldn't be work for us, why the diamond drills shouldn't be set up to test our finds at depth, and thus answer the big question, did our claims have the makings of a mine?

Every time an airplane was heard or actually landed near our camp our hopes rose, that Pat might finally return, perhaps even bringing the big boss with him. However, as so often before, we were in for a long period of waiting. Senseless, disheartening waiting, wasting days and weeks when every hour should have counted. True, we kept working, prospecting our claims and the immediate neighborhood. But there was so much unexplored ground left. What were we doing? We were hanging around staked, recorded, and tagged claims, secured by assessment work that was sufficient for years.

Who could tell where the greatest treasures might be hidden? It had to be found and there was only one way to do it, to keep on exploring, and that, after all, was the only satisfying activity for anyone who had in him the stuff that made a good prospector. He was a restless fellow, who kept going by himself, to whom the far hills always held a promise, and an irresistible lure. Avoid clipping his wings, allow him to roam, and he would find the stuff, stake the claims, and move on. There were few of his type, but plenty of others who prefer to work in a gang, in a big camp.

Let those who preferred that big camp follow the prospector to do the digging, trenching, drilling, and blasting while the restless seeker headed for a new hunting ground. That had always been my contention, and many a time I had expounded my ideas to Pat and my chums like Art, Ronny, and Steve, all three of whom shared my views.

Given more freedom of action, we could have made a few good teams of trailblazers. Instead, we felt stymied and tied down. And yet, the days passed, the long, hot days of the short northern summer.

The picture my memory recalled of those days of waiting seemed somewhat hazy. Not much happened to distinguish one day from the other. It seemed as if the smoky haze from innumerable fires that ravaged the dried-up country gave the whole period its character. Not before the latter part of August did word come that the big boss and Pat were on their way in.

"We'll believe that when we see them," we all said, but in spite of the lack of optimism, we spruced up our camp. Only the sight of the trim, new Beechcraft of Mackenzie Air Service circling over our camp one morning convinced us that the time for new action had come.

Once the big boss, Pat and another mining expert had disembarked, we immediately began a tour of our high-grade showings, keeping the glory hole to the last as our trump card.

For two days we tramped the well-worn trails with our visitors, but only after the last stop, where the big nuggets were gleaming did the boss finally open up.

"Your shows look good," he said, "but only the drills can tell the whole story. Where do you figure on putting up your winter camp?"

"Just over the next ridge! There we have a sandy plateau with plenty of dead but sound timber for our buildings and for fast and easy access to the lake. It's an ideal set-up. We have studied it for a long time," I said.

"Then you were pretty sure?"

"No, sir, not so sure but hopeful."

"Well, you have to be, or you wouldn't be a prospector! Have you got any good samples for me to take out? They tell more than words."

"The ones we have will talk all right!"

In high spirits we returned to camp, and a happy company gathered for supper. The big boss and his friend retired early, they were to fly out early the next morning, but Pat and I talked deep into the night about the necessary preparations for diamond drilling, the building to be put up and supplies to be brought in without delay. Time was of the essence. Another month and the snow might fly.

Suddenly the spirit of the old days had come alive again, the "it will take some hustling, but we will do it!"

With the dawn of a new day, we were up. As usual, I busied myself to look after the feeding of the gang, trying to make it a blue plate special in honor of the occasion and of our guests. The happy and hungry gang did the breakfast full justice.

Over our coffee and smokes we sat and talked. Somebody asked the big boss what he thought of the political situation. But we weren't too serious about it. In the North these things always seemed somehow far away, and especially on this bright morning, it was so much more important that we were going to build a winter camp, that the diamond drills would soon be doing their stuff, and as for myself, I was thinking that soon my little family might be living in a log-cabin. Little did we know.

A few minutes later all was changed. The Beechcraft roared over our tents, landed crosswind none too smoothly, and came racing toward camp. "What's the matter with that guy this morning?" Pat squeezed out.

Grave of manner, the usually debonair Harry Hayter stepped ashore, looked around at us and quietly, too quietly, said, "'It's war. The Nazis marched into Poland!"

Our dreams were completely shattered. For a long time, not a word was said. Then the boss turned to us, "Sorry, boys! This changes everything. Better get ready to pull out. If it blows over, and we must hope it will, we'll meet again! So long for now, and good luck to all of you!"

Silently we shook hands. Our guests went aboard.

We pushed the plane away from the dock. Harry waved to us for the last time, gunned his motor and skimmed away over the morning's still water to rise quickly and turn south.

Silent and sad we watched the air ship disappear, and with it our dreams. The sound of the motor died away. Then we were all alone. For a long, long while we sat near the dock, each one busy with his own thoughts, trying to look into a dark future, which only an hour or so ago was shining with golden promise.

I think it was old Fred who finally roused us from our somber brooding.

"Come on, fellows! No use being downhearted. Might as well be merry. You may soon be in the trenches!"

Of course, he was right, and there was nothing else left, but to pack up and pull out. Not much was done though, that day. It was not so easy to part with one's hopes and dreams. Slowly and reluctantly a rather pensive bunch of fellows started on the last task of building a bigger and higher cache, to hold all the equipment safely until we might return.

When the plane came for us a few gloomy aimless days later, we folded our tents and rolled up our eiderdowns for the last time.

Goodbye to Pensive Lake!

Goodbye to the North!

So ended my story about the North, not much of an ending, but that might be right. Someday there might be a new beginning. As old Fred, who stayed behind in Yellowknife said, when we shook his rough fist for the last time, "I'll stick around. I'll wait till you guys get back!"

Time flew. Fred would not be waiting any more, but the North will be forever there. As long as the northern lights waft their magic, mysterious veils across the skies of an evening in spring and I heard the call of the northward flying geese from high above, I would not forget.

Someday, I thought, I must return. "It gets you" they say down North.

Would you want to have come North with me?

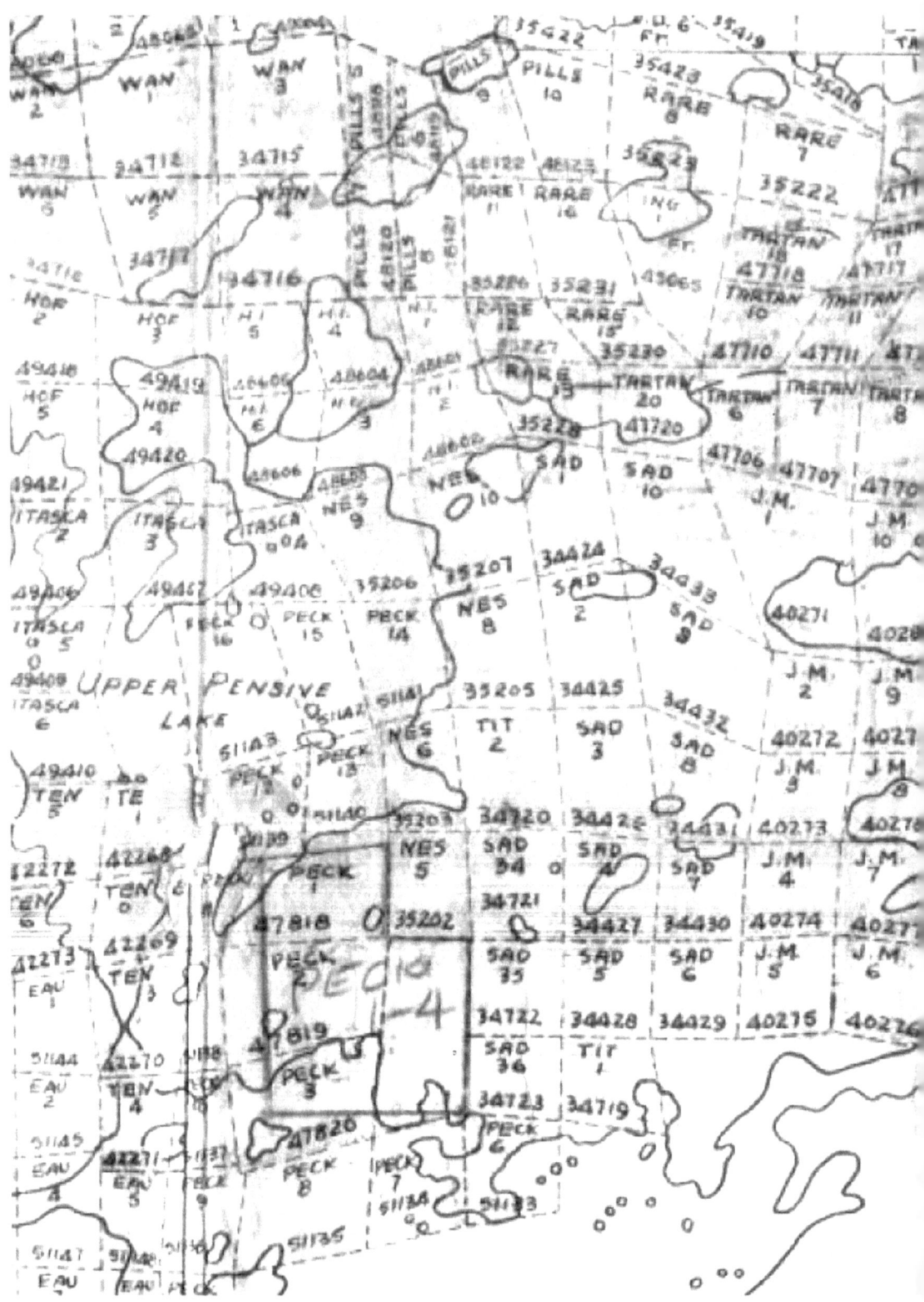

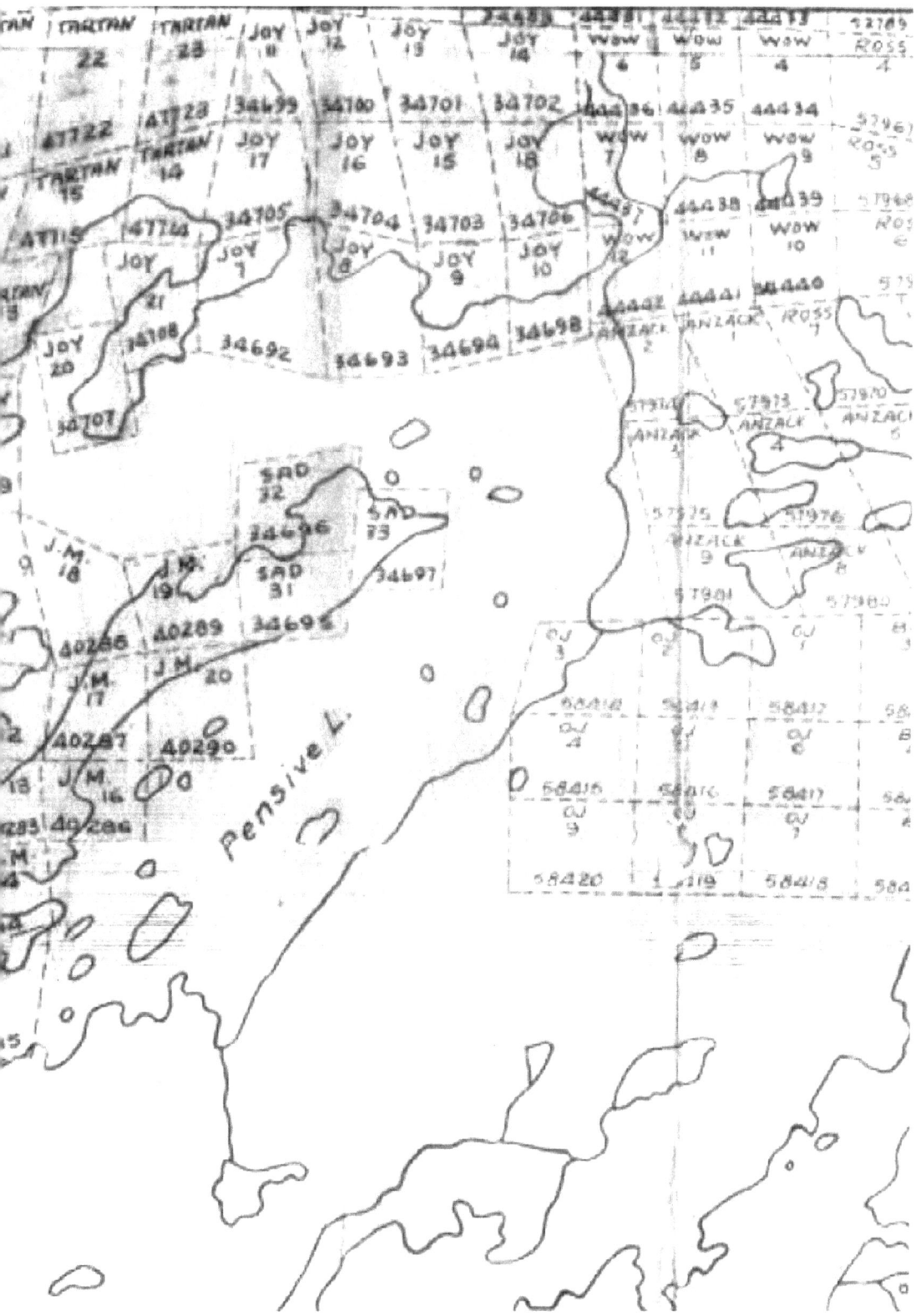

Printed in Canada